Studies in Computational Intelligence   431

Editor-in-Chief

Prof. Janusz Kacprzyk
Systems Research Institute
Polish Academy of Sciences
ul. Newelska 6
01-447 Warsaw
Poland
E-mail: kacprzyk@ibspan.waw.pl

For further volumes:
http://www.springer.com/series/7092

Wei Ding, He Jiang, Moonis Ali,
and Mingchu Li (Eds.)

Modern Advances in Intelligent Systems and Tools

Editors
Wei Ding
Department of Computer Science
University of Massachusetts Boston
Boston
USA

Moonis Ali
Department of Computer Science
Texas State University San Marcos
San Marcos
USA

He Jiang
School of Software
Dalian University of Technology
Dalian
China

Mingchu Li
School of Software
Dalian University of Technology
Dalian
China

ISSN 1860-949X
ISBN 978-3-642-42899-9
DOI 10.1007/978-3-642-30732-4
Springer Heidelberg New York Dordrecht London

e-ISSN 1860-9503
ISBN 978-3-642-30732-4 (eBook)

Preface

Artificial intelligence has become a popular research field with wide applications. In industrial, engineering, and other applications, artificial intelligence is applied to construct intelligent systems for solving real-world problems. Based on the development of artificial intelligence methods, modern intelligent systems have played an important role in many areas, including science, industry, engineering, automation, robotics, business, finance, medicine, and cyberspace. Meanwhile, a great number of intelligent systems, from prototypes to real-world tools, are proposed to cope with new problems. Intelligent systems provide a platform to connect the research in artificial intelligence to real-world problem solving applications.

In solving real world problems, intelligent systems and tools employ many techniques from areas such as machine learning, data mining, social networks analysis, natural language processing, and cognitive computing. Various intelligent systems have been developed to face real-world applications. Intelligent healthcare systems, decision making systems, design, planning, and scheduling systems, and human-computer interaction systems are just a few examples. The modern advances in intelligent systems and tools present research results in areas which represent the frontiers of applied artificial intelligence.

This book is published in the "Studies in Computational Intelligence" series by Springer-Verlag. It discusses the modern advances in intelligent systems and the tools in applied artificial intelligence. It consists of twenty-three chapters authored by participants of the 25th International Conference on Industrial, Engineering & Other Applications of Applied Intelligent Systems (IEA/AIE 2012) which was held in Dalian, China. The conference followed the IEA/AIE tradition of providing an international scientific forum for researchers in the diverse fields of applied intelligence. The conference was organized by School of Software, Dalian University of Technology and was sponsored by the International Society of Applied Intelligence (ISAI).

The book is divided into six parts, including Applied Intelligence, Cognitive Computing and Affective Computing, Data Mining and Intelligent Systems, Decision Support Systems, Machine Learning, and Natural Language Processing. Each part includes three to five chapters. In these chapters, many approaches, applications, restrictions, and discussions are presented. The material of each chapter is self-contained and was reviewed by at least two anonymous referees to assure the high quality. Readers can select any individual chapter based on their research interests without the need of reading other chapters. We hope that this book provides useful reference values to researchers and students in the field of applied intelligence. We also hope that readers will find opportunities and recognize challenges through the papers presented in this book.

We would like to thank Springer Verlag, Professor Janusz Kacprzyk, the editor-in-chief of this series, for his help in publishing the book. We also would like to thank Professors Zhongxuan Luo, Xindong Wu, Nick Cercone, and Ramamohanarao Kotagiri for their advice. We greatly thank the reviewers for their hard work in assuring the high quality of this book. Besides, a grateful appreciation is expressed to Jifeng Xuan, Zhilei Ren, and Runqing Wang for their help that made the book possible. We thank Dr. Thomas Ditzinger, the senior editor of Springer-Verlag for his hard work in preparing and publishing this book. Finally, we cordially thank all the authors who did important contributions to the conference and the book. Without their efforts, this book could not be published successfully.

April 2012

Wei Ding
He Jiang
Moonis Ali
Mingchu Li

Contents

Data Mining and Intelligent Systems

Decision Support Systems

Machine Learning

Natural Language Processing

Applied Intelligence

A Non-Contact Health Monitoring System Based on the Internet of Things and Evidential Reasoning

Xingli Zhao, Hong Zhang, Yajuan Zhang, and Ning Yang

School of Computer and Information Science, Southwest University Chongqing, China
zhangh@swu.edu.cn

Abstract. With the development of science and technology, the Internet of things has become a hot spot and is used in various fields. In this paper, a new model which is called A Non-contact Health Monitoring System Based on the Internet of Things and evidential reasoning is proposed to assist practitioners in the assessment of patents' security. The processing methods used in the monitoring system are Analytical Hierarchy Process and Dempster–Shafer theory. The System brought a great of convenience to the families and also provided a healthy guarantee for users.

Keywords: the Internet of Things, Non-contact Health Monitoring System, Analytical Hierarchical Process, Dempster–Shafer theory.

1 Introduction

As the trends of the average life expectancy, the decline of the birth rate and health-oriented are increasingly clear, people demand better health care system. As a result, realizations of a non-contact health monitoring system (NCHMS) become more vital and urgent. And the Internet of Things technology will be the best means of its implementations. The Internet of Things not only provides the health care system for aspects of technical support, but also will solve the core problems--collecting and processing the remote medical information. The system refers to an information system that can remotely monitor the user's physical condition and physiological parameters. In this paper, a remote health care program design based on the Internet of Things is provided and it can improve intelligence of information collection and process, which will promote applications of the Internet of Things in the medical industry.

In this paper, Section 2 begins with a detail introduction to the Non-contact Health Monitoring System based on the Internet of Things. Then, Section 3 introduced the procedure of the proposed model for evaluating the patients' status risk. Section 4 concludes the paper.

2 The Non-Contact Health Monitoring System (NCHMS)

The idea of this paper stems from the concept of Interne of Things. When the concept of Internet of things is introduced to the smart home, its scope has been expanded more,

W. Ding et al. (Eds.): Modern Advances in Intelligent Systems and Tools, SCI 431, pp. 3–8.

including home security, home health care, home entertainment, etc. Because not every family has the ability to own a monitoring instrument, we put forward the non-contact health monitoring system, which is composed of common digital equipment, such as cameras, environment detector, digital cameras etc. The system uses these devices to collect the user's facial expressions and posture pictures, takes advantage of microphones and other equipment to collect the user's voice. When the patient appears painful, sad or other facial expressions, the monitoring system will analyze and combine all the data, then timely identify the patient who is risk or not. Then the system will remind the user and send the information to the hospital.   If necessary, the hospital will inform the nursing staff and the family members to take appropriate measures. Fig.1 shows the overall structure of the non-contact health monitoring system.

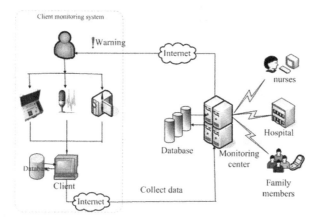

Fig. 1. The non-contact health monitoring system（NCHMS）

The monitoring system mainly consists of two parts, the client and the server. Client is responsible for the collection of user's facial expressions and postures data, as well as pretreatment. The server center is mainly responsible for data analysis, management and processing.   The client and the server are connected through the Internet. On the client, user's data are collected through the cameras, digital cameras, microphones and other equipment. The collected data are classified and stored in different databases. Then the client pre-processes the data through its own procedures and identifies the user's current facial expressions, postures and sounds. In the system, we consider the external factors, such as temperature, air quality, weather and so on. In addition to these factors, the patients' habits are also included, so the Non-contact Health Care System based on the internet of things and information fusion is not a universal system, it is a personalized monitoring system. That is to say, we choose different parameters for the different characteristics of the patients, then analyze the patient data, and integrate all the considered factors with the evidence theory. After these, the final status of the patient's condition will be given. And based on the result, the workers choose the different selection for the patient.

3   The Development of Proposed Model

The purpose of this study is to design and develop a new model which can help with the assessment of the patients' risk. The construction of the model is divided into six phases, as depicted in Fig. 2. The first step is to collect the data, Then the following two phases is AHP [3] section, in which AHP is applied to construct a hierarchical structure of patients' risk and derive the weights of each issue presented in the structure. In the rest three phases, evidential reasoning has been used to combine all the evidences together and transform the combination result into probability distribution to get a consensus decision for patients' risk. Detailed descriptions of each phase are presented in the following sections.

Phase 1: establishing a hierarchical structure
The first step is to establish a hierarchical structure of the problem. The problem, which is evaluating the degree of the patients' risk, is classified into four major levels, as depicted in Fig. 3. The first level of the hierarchy is the goal of the assessment: the degree of the patients' risk. Level 2 represents two major aspects influencing the risk. In level 3, the external factors are divided into 3 smaller factors. The inertial factors are divided into four smaller factors. And the ultimate variables associated with the patients' risk are contained in level 4.

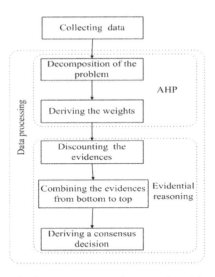

Fig. 2. Development of proposed model

Phase 2: constructing pairwise comparison matrix
Once the structure is established, the weight of each items need to be derived. All the weights are calculated through a process called pairwise comparison.

Example 3.1. Assuming variable V_2 is strongly important than V_1, the score of importance between them would be 3. Therefore, the element m_{21} in their pairwise comparison matrix, which represents the relative importance of V_2 over V_1 can be derived: $m_{21}=3$. And the value of m_{12} can be calculated as: $m_{12} = \dfrac{1}{m_{21}} = \dfrac{1}{3}$.

Thus, according to the scores of importance between each variable given by experts, we can construct the pairwise comparison matrix. Then calculate the final weights of each variable with the AHP [2].

Phase 3: preparing evidences for combination
With the weights of each variable, Dempster–Shafer theory of evidence [2] will be applied. Consider the frame of discernment with 3 hypotheses $\Theta=$ {Low (L), Medium (M), High (H) }, which describe the degree of patients' risk. Variables with its descriptions using the 3 hypotheses are regarded as evidences in the DS theory. However, these evidences are preliminary, which are too vague to be combined together directly. Therefore, before we apply Dempster's rule of combination, it is needed to give all the preliminary evidences a discount by their confidential level, which is defined as follows:

Definition 3.1. Suppose only two adjacent fuzzy variables have the overlap of meanings and the intersection of them is not empty. Given a confidential level α of the evidence that the evidence is reliable, then the mass function m' on Θ becomes:

$$m'(A) = \alpha\, m(A), \forall A \in \Theta, A \neq \Theta$$
$$m'(Y, A) = \frac{S(Y \cap A)}{S(X \cap A)}(1 - \alpha\, m(A)), Y \neq A, Y \in X, X \subset \Theta$$

here X is the set of elements whose intersection with A is not null. $S(Y \cap A)$ denotes the intersection area of Y and A, while $S(X \cap A)$ denotes the total intersection area of all the elements in X with A.

In this step, we use the Fuzzy sets. Fig.3.shows the membership functions based on our proposed model (NCHMS).

Definition 3.2. Let X be the universe of discourse, which include three linguistic variables describing the degree of risk, X = {Low, Medium, High}, assuming that only two adjacent linguistic variables have the overlap of meanings. And let A be a fuzzy set of the universe of discourse X subjectively defined as follow:

$$f_{low}(x) = \begin{cases} 1, & 0 \leq x \leq 5 \\ -0.1x + 1.5, & 5 \leq x \leq 15 \end{cases}$$

$$f_{medium}(x) = \begin{cases} 0.1x - 0.5, & 5 \leq x \leq 15 \\ -0.1x + 2.5, & 15 \leq x \leq 25 \end{cases}$$

$$f_{high}(x) = \begin{cases} 0.1x - 1.5, & 15 \leq x \leq 25 \\ 1, & 25 \leq x \leq 30 \end{cases}$$

where $f_{high}(x)$, $f_{medium}(x)$, $f_{low}(x)$, , are the membership functions of the fuzzy sets, which are shown in Fig. 4. Then the weights derived in Phase 3 can be regarded as the confidential level α of each evidence.

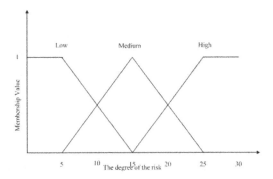

Fig. 3. Membership function of security

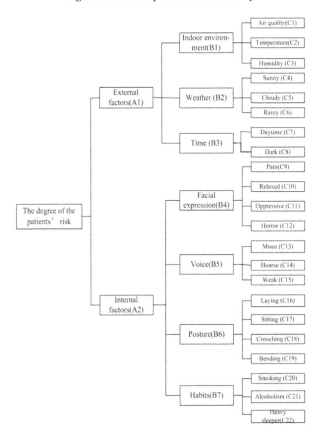

Fig. 4. Hierarchical model of the patients' risk

Phase 4: Combining the evidences from bottom to top.
Once the discounting is accomplished, we can use Dempster's rule of combination to combine all the evidences from bottom to top. However, the results from the combination rule of DS theory often produce results which do not reflect the actual distribution of beliefs. In order to handle that, averaging is integrated into the combining rule. If there are n pieces of evidences, the masses assigned to the same set should be averaged, which avoids overdependence on a single piece of conflicting evidence. Then, use the fusion formula to combine the evidence with itself (n-1) times. Detailed process of this method is given in [5].

Phase 5: Deriving a consensus decision.
Having combined all the available evidences from bottom to top, we then turn to the problem of making a decision. To do this, we can adopt the pignistic probability transformation as introduced [5,4], which is applied to convert a belief function to a probability function. After the transformation, each variable with its probability is then ranked by calculated pignistic function scores. The one with the highest score is our decision.

4   Conclusion

The thought of the Internet of things greatly changes the life of people. Remote health monitoring, remote diagnosis, family physicians and other convenient efficient and effective methods were born as a result. The idea of this paper is based on the background. According to the features of NCHSM, this paper has proposed a more refined assessment model and assessment method for the patients' dangerous. Although the assessing variables may be affected by contingent variables of different circumstances, the proposed model can easily be modified to those variables. It is believed that this model will be helpful in the assessment of the patients' risk.

Acknowledgments. This work is supported by the Fundamental Research Funds for the Central Universities (XDJk2009C018).

References

1. Chen, M.K., Wang, S.C.: The critical factors of success for information service industry in developing international market: Using analytic hierarchy process (AHP) approach. Expert Systems with Application 37, 694–704 (2010)
2. Shafer, G.: A Mathematical Theory of Evidence. Princeton University Press, Princeton (1976)
3. Saaty, T.L.: How to make a decision: The analytic hierarchy process. European Journal of Operational Research 48, 9–26 (1990)
4. Deng, Y., Jiang, W., Sadiq, R.: Modeling contaminant intrusion in water distribution networks: A new similarity-based DST method. Expert Systems with Applications 38, 571–578 (2011)
5. Zhang, Y.J., Deng, X.Y., Jiang, W., Deng, Y.: Assessment of E-Commerce security using AHP and evidential reasoning. Expert System with Application 09 (2011)

Real-Time Alarm Management System for Emergency Situations

Ana Cristina Bicharra García[1], Luiz Andre Portes[1], Fernando Pinto[1], and Nayat Sanchez-Pi[2]

[1] Computer Science Institute,
Fluminense Federal University.
Rua Passo da Pátria, 156. Niterói. RJ. 24210-240. Brazil
{cristina,luis.andre,fernando}@addlabs.uff.br
[2] Computer Science Department,
Carlos III University of Madrid.
Avda de la Universidad Carlos III, 22. Madrid. 28270. Spain
nayat.sanchez@uc3m.es

Abstract. In this paper, we present an alarm management system focused on guiding offshore platform operators' attention to the essential information that calls for immediate action during emergency situations. Due to the imminent associated danger involved in the petroleum operation domain, only well trained workers are allowed to operate in offshore oil process plants. Although their vast experience, human errors may happen during emergency situations as a result of the overwhelmed amount of information generated by a great deal of triggered alarms. Alarm devices have become very cheap leading petroleum equipment manufacturers to overuse them transferring safety responsibility to operators. Not rarely, accident reports cite poor operators' understanding of the actual plant status due to too many active alarms. A petroleum process plant can be understood as a system composed of a set of equipments interacting with each other to transform and conduct safely a fluid. Each equipment has its own set of rules and safety devices (alarms). The system is subjected to external, non-predictable, effects coming from nature. Hence, the petroleum process plant system can be represented as a set of agents with rules for acting, reacting and interacting with each other. Each equipment is represented as an agent. This AI multi-agent based approach is the basis of our alarm management system for assisting operators to make sense of alarm avalanche scenarios. Our model was implemented using stored procedure statements, installed into the automation circuit of a actual offshore petroleum platform and we are currently collecting results. During initial tests we identified unexpected benefits concerning verification of the process plant automation procedure.

Keywords: emergency situations, alarm management, multi-agent systems, oil industry.

W. Ding et al. (Eds.): Modern Advances in Intelligent Systems and Tools, SCI 431, pp. 9–18.
springerlink.com   © Springer-Verlag Berlin Heidelberg 2012

1   Introduction

Alarm management in emergency scenarios has become a topic of great concern in different economic sectors, such as nuclear, aeronautics and offshore oil industry, due to the frequent accidents occurred in the last decades caused by inappropriate alarm management systems. Although great effort has been devoted to plant's automation and cheap alarm device development, operators play an important role mastering all information and adjusting equipments' behavior as needed. The observations of our research are domain independent, but, in this paper, we focus on the offshore oil process plant domain.

An alarm is an informational representation of an equipment or system abnormal situation. As sensor devices became inexpensive and simple to embed into equipments, manufacturers overused installing sensors into their equipments, so to better comply to safety norms. An alarm informs operators of the process plant unit' status and might require immediate action. Trained operators handle pretty well few, but not many, alarm information at a time. During a non-planned process plant shutdowns, operators face an avalanche of alarm information, frequently over 1000 alarms/minute, that must be understood, prioritized and reasoned to decide upon which action, if any, to take.

This information overflow has been related as one of the major cause of several serious accidents in the last decade, such as the Mildford Haven refinery accident in the UK, on 24 July 1994, which resulted in a loss of 48 million pounds and two months of non-operation. The report of the Health and Safety Executive department [1] has identified as the accident cause the refinery operators' inability to identify what was really happening behind the large number of triggered alarms generated. The accident could be avoided if the alarm system had identified the cause of the problem, sorted alarm information and displayed only the most important ones to the operator be able to act properly.

A process plant of a petroleum offshore unit is a complex artifact composed of independent equipments which interact with each other to receive petroleum from subsea reservoirs; treat it and export gas and oil to land refineries. Each equipment behaves and reacts to accomplish its own goal, such as compress gas or treat oil, but maintaining its behavior within safety functioning range. Sensors and actuators are essential devices embedded in the equipments to monitor or control their behavior, respectively. These complementary devices are orchestrated by the process plant automation system that receives sensors information and triggers actuators actions, such as closing a valve. Moreover, automation systems also send operators important information concerning the process plant status.

Inspired by the distributed and encapsulated aspect of the process plant artifact physical model, we proposed a multi-agent-based alarm management system to synthesize the process plant situation during emergency situations. Each agent represents an equipment that understands about its expected and unexpected behavior within the process plant. During emergency scenarios, alarms related to expected behavior can be suppressed to lead operators' attention to unexpected behavior. Distinguishing expected from non-expected behavior during emergency scenarios and using this information to filter what to display to operators is the basis for our intelligent alarm management system proposal. In addition to proposing a model, we have implemented a prototype version and tested it in a controlled environment. We are currently deploying a version of the system to work within a Brazilian offshore petroleum process plant unit. The rest of the paper is structured as followed. First section presents related work in the area of alarm management systems, focused on offshore oil process plant domain. Next, our multi-agent alarm management system is laid down, followed by a case study. Results are presented and a conclusions are outlined.

2 Related Work

In 2000 there were at least eight oil spills in Brazil. It is estimated that the sum of fines imposed to oil companies, as a result of environmental stresses, along that year, exceed two hundred million dollars [2]. Data from the Federation of Oil (FUP) [3] and the National Front of Oil (PNF) [4] show that since 1967 there were more than 350 accidents involving fatalities in petroleum industry . In an attempt to ensure proper operating environment for maximum security for people, environmental protection and operational continuity, leading enterprises have been working steadily in their policies of SMS (Safety, Environment and Health). On the other hand, the world economic growth has called for an oil production increase.

Process plant automation and alarm management systems are a reality in both as academic proposals and on-the-shelf technology. Some of them are:

- The Matrikon Alarm Manager software [5] that allows multiple tests to be made for an alarm system. It provides operators with reports in Microsoft Excel spreadsheets or accessed via web browser. The material can be used to study the artifact's behavior using basic statistic functions to access alarms' information and advanced analyzes that include a detailed mapping of the condition of the alarm system, focusing on typical problems that can be detected from statistical tools.
- ABB [6] delivered an alarm management feature as part of an automation system.
- GEs Alarm Response Management system [7] that evaluates the response of each individual alarm to the device's condition while also comparing responses to multiple alarms using the critical statistic is Mean Time To Repair (MTTR) to identify trends.

Most alarm management approaches focus either on post-crisis information analysis or on more automation procedures. None addresses the challenge of assisting operators to make sense of the situation DURING crisis scenarios. Since our goal is to provide assistance during crisis, our system must be designed as a real-time application. We investigated many different approaches such as conventional centralized structures, decentralized applications and mult-agent system. The last has presented the best results concerning response time and flexibility to grow the application.

3   MAS for Alarm Management System

An emergency shut down (SD) in a process plant unit is a safety measure, in general, triggered by automation systems (AS) to protect the equipment, the system, people operating the system and the environment. Each SD triggers a set of events designed by the automation designers to protect the unit. For example, during a shut down of a specific equipment, it is expected this equipment to be contained and isolated from the rest. For accomplishing these effects, the automation procedure imposes the closing of upstream and downstream valves. When shutting down a system or even the entire unit, many more events should be triggered. Of course, these events interact with each other and further automation procedures must specify the desired interactions. At the same time sensors monitor parameters' values for undesirable situations, sometimes this undesirable situations are the expected ones. For instance, a fast pressure drop downstream a pump my represent a pump cavity danger. However, the same information when associated with a pump turn-off means a correct and desirable behavior. As noted, in an emergency situation, a great deal of information is generated. It is extremely useful to establish priorities to set which alarms are actually important to be displayed to operators.

Agents technology perfectly fits in this kind of engineering problems due to the main characteristics of agents: autonomous reasoning, proactivity, communication and adaptive behavior. During the years researchers have come to the conclusion that reactivity is also a very important characteristic that an intelligent agent should possess [8]. Reactivity is suitable for dynamically changing environments performing an immediate response to some changes which have been recognized and perceived [9]. Stored procedures also presented as a mature technology, used in many different complex domains, that must run within the automation process plant network.

The objective of our approach is to work as an alarm information filter, receiving and sorting information sent by the automation supervisor system (AS), during a serious non-planned process plant shutdown of level ESD-2, ESD-3P or ESD-3T, called here simply as STOP. This STOP situation causes an avalanche of alarm information. It is humanly impossible for process plant operators to understand not only what is happening with the process plant, but also, and more importantly, if it is being moving to a safe state, i.e. if the plant os properly being turned off . Thus, our proposed system can be seen as a assisted-stop system. The operator should only receive information related to unexpected alarms or danger degree escalation that may compromise the safety of the unit.

Each equipment has its own rules of appropriate and inappropriate behavior associated with normal and crisis scenarios. Automation systems receive sensors' information of each equipment and send to operators a list of alarms they should pay attention to. In a crisis, automation systems has a built-in set of events related to actuators proper actions that must be triggered. Distributed agents that understand expected and expected behavior related to equipment behavior and equipment's reaction to automation events are the basis of our multi-agent approach described next.

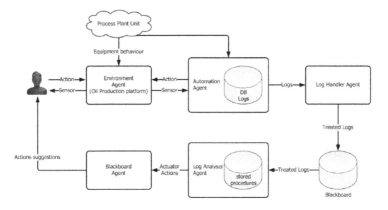

Fig. 1. Multi-Agent Architecture

3.1 MAS Architecture

Agents paradigm provide an excellent modeling abstraction for our intelligent alarm management system due to agent's human-like characteristics, including reasoning, proactivity, communication and adaptive behavior. Moreover, Alarm Management Systems in Emergency Situations beg for technologies that are transparent, so that the functional behavior in an emergency situation can be easily understood by operators.

Our model, illustrated in Figure 1, represents an intelligent support system for alarms management in an offshore oil production domain. MAS architecture is composed of four types of agents: Environment Agent, Automation Agent, Log Handler Agent, Log Analyser Agent and Blackboard Agent. The agents main functionalities are the following:

- **Environment Agent:** This agent monitors information from nature. It also manages information regarding the Oil Production Platform status such us the identification of a SD.
- **Automation Agent:** Automation System (AS). Events and alarms continuously monitored and identified by sensors embedded in the equipments are sent to the automation system (AS). Later, the AS triggers Actuators (pumps, valves, compressors, etc.) for actions (open/close, turn-on/turn-off). This agent creates a log of events (log history) merging the information coming from sensors to actions sent to actuators. This log is recorded in the Blackboard, accessible by all agents.
- **Log Handler Agent:** This agent reads and parsers the information log in the blackboard to created a structured information that can be further analyzed. The structured information is also registered in the Blackboard.
- **Log Analyser Agent:** This agent is actually a set of agents implemented as stored procedures. Each agents understands about a equipment. Each agent selects, from the structured information stored in the Blackboard, only the information that concerns its expertise. Its knowledge is written in terms of

production rules describing expected and unexpected behaviors. Expected behaviors triggers a alarm information suppression action, that means an information removal from the blackboard.

– **Blackboard Agent:** This is agent handles information that will be displayed to operators. It must handle information synchronization since many agents are reading and writing into its structure. It invokes the GUI where alarms information are shown to the final operator.

3.2 Ontology

We model the process plant domain using an ontology that emphasizes the component and monitoring characteristics of the artifact. Modeling this environment involves representing entities and relationships among these entities. The main concepts of the ontology and its description are illustrated in Figure 2. The domain main entities involve:

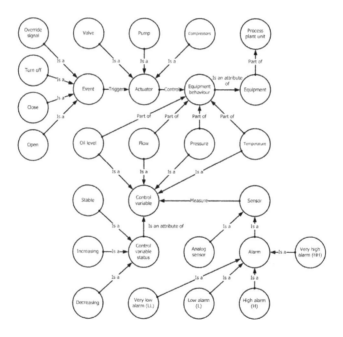

Fig. 2. Ontology for Alarm Management System

– **Equipment:** It is a component of a process plant unit.
– **Actuator:** It is a device, such as valves, pumps and compressors, which controls the equipment behavior.
– **Equipment behavior:** It represents the way the equipment behaves in order to achieve a desired functionality. Equipment behavior is composed of oil level, flow, pressure and temperature. Any behavior is considered a control variable which is measured through sensors.

- **Event:** It is an action over the actuators that might causes a change in the state of the alarm. For instance, the event "close" over the actuator "valve" should cause a decrease on oil flow in the pipeline.
- **Alarm:** It represents an abnormal state of the equipment behavior. The possible values are High (H), Very High (HH), Low (L) or Very Low (LL). HH and LL leads to equipment and even the entire unit shut down
- **Sensor:** It is a device that measures control variables. There are two type of sensors: Analog sensors and Alarm sensors. Analog sensors measure the exact value of the variable such as: oil temperature $= 40\,°C$. Alarm represent discrete sensor information that states danger situations.
- **Control Variable Status:** It indicates the variation of a measurement. A control variable status indicates for instance if the temperature is increasing.

An equipment is part of a process plant unit. Equipment achieves its goals though a set of behaviors such as oil level, pressure, temperature and flow that are monitored by sensors. Alarm is a special type of sensor that indicates an equipment behavior overtake danger threshold values. There are four types of alarms: Very high, high, low and very low alarm. There is also analogic sensors that measure the exact value of a given behavior. Automation controls equipment behaviors though events changing actuators status such as pump turn-off. Valves, pumps and compressors are examples of actuators.

3.3   Stored Procedures

All agents reasoning are represented as production rules implemented in terms of stored procedures. Each agent has a set of stored procedures that works autonomously as soon as it detects a change on the environment. Agent's environment differs. The Log Handler environment considers its environment the events' log containing the events written by the automation system.

The Log Handler agent reasoning contains rules to parser the information sent by the automation system. It uses a lexicon containing a list of commands it is able to recognize and a syntax to guide its message interpretation. This agent is responsible for the first level of information filtering. The log analyzer agents keep monitoring the Blackboard containing the treated log messages. From time to time (it has a configurable clock), it reads the changes and checks whether changes follows the expected behavior. Each agent uses a set of specific rules of good and bad behavior defining a specific equipment. Last, but not least, the interface agent monitors the Blackboard agent to check whether there is a change in the operators' information interface that must be done.

The rules for all agents are fed by automation experts though a special knowledge acquisition interface (not included in the main model) and stored in the knowledge-base. There are two types of rules: those depending only on the status of each equipment behavior and those that depends on a combination of equipments' behaviors. Therefore, there is a rule chain.

As named, stored procedures are programs stored in a database. Relational programming are programs developed using sequences of SQL statements. Due to the extremely efficiency of database engines on processing of SQL and the physical proximity of the stored procedures with data, the strategy was the best choice and the results have proven it so far. Follow a sample of a procedure used by the Log Analyzer agent.

```
PROCEDURE LOG_ANALYSER( ) IS ... BEGIN ...

INSERT INTO T_PARADAS_EM_ANDAMENTO
(TIPA_CD_IDENTIFICACAO,TIPA_SG_TIPO_PARADA,PARA_DT_INICIO,OCAL_CD_IDENTIFICACAO)
(SELECT
      P.TIPA_CD_IDENTIFICACAO,
      TP.TIPA_SG_TIPO_PARADA,
      MIN(P.PARA_DT_INICIO),
      MIN(P.OCAL_CD_IDENTIFICACAO)
FROM
      (SELECT DISTINCT
             PARA_CD_IDENTIFICACAO,
             TIPA_CD_IDENTIFICACAO,
             PARA_DT_INICIO,
             OCAL_CD_IDENTIFICACAO
      FROM PARADA
  START WITH (PARA_DT_INICIO <= DT
                     AND PARA_DT_FIM IS NULL)
      CONNECT BY NOCYCLE (PARA_DT_INICIO <= DT
                     AND (PARA_DT_INICIO
                                  BETWEEN PRIOR PARA_DT_INICIO AND
                                  NVL( PRIOR PARA_DT_FIM, DT)
  OR PARA_DT_FIM + V_TOLERANCIA
       BETWEEN PRIOR PARA_DT_INICIO AND NVL(PRIOR PARA_DT_FIM, DT)))) P
      INNER JOIN TIPO_PARADA TP ON P.TIPA_CD_IDENTIFICACAO = TP.TIPA_CD_IDENTIFICACAO
      GROUP BY
             P.TIPA_CD_IDENTIFICACAO,
             TP.TIPA_SG_TIPO_PARADA);
```

4   Case Study

Log analyzer periodically checks incoming data in order to detect ESD occurrences. When they are detected the analyzer starts evaluating rules in a forward chaining way. When rules evaluation concludes a event failure, the event description and a set of related alarms are sent to the blackboard to be displayed to operators. The alarms not related to failed events are suppressed from the blackboard, what relieves operators from dealing with the complete and large sets of alarms. The following presents an example of rules for evaluating events.

Rule 1: *if alarm TAHH123324A is active OR alarm TAH123323A is active then a wrong behavior of the serpentine of the reboiler P-Z-123301-02A is detected*

Rule 2: *if the alarm TAHH123324A is inactive and alarm TAHH123323A is inactive then an expected behavior of the serpentine of the reboiler P-Z-12330102A is detected*

Rule 3: *if a wrong behavior of the serpentine of the reboiler P-Z-123301-02A is detected and no expected behavior of the serpentine of the reboiler P-Z-12330102A is detected the event Turn off the serpentine of reboiler P-Z-123301-02A is failed! then the event Turn off the serpentine of reboiler P-Z-123301-02A is failed!*

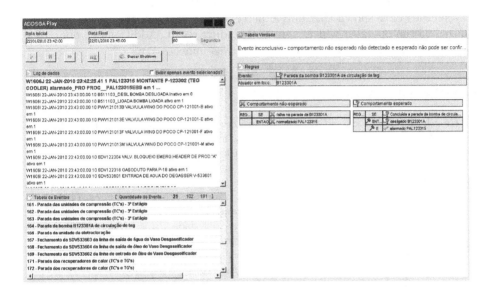

Fig. 3. Logs, Events and Rules Relationship

Figure3 shows the results of the analyses of data from 9 SDs. The last column represents the percentage of alarm suppression. As shown, there are scenarios of 93,76 percent of suppression. After executing our systems in 19138 logs, taking 517694 milliseconds, the results were: 349 triggered events and 664 of triggered rules. The total amount of information displayed to operators is displyaed in Figure 4.

SHUT DOWN EXECUTION					ALARMS			
ESD	Date	ESD type	Begin	End	Total	Suppressed	Presented	% Suppresed
1	22/1/10	ESD-2	23:42:25:40	23:44:01:26	71	65	6	91,55%
2	22/1/10	ESD-2	23:45:36:06	23:47:59:22	58	52	6	89,66%
3	22/1/10	ESD-2	23:50:22:40	23:52:46:05	33	33	0	100,00%
4	30/3/10	ESD-2	13:34:15.84	13:35:59.34	27	27	0	100,00%
5	30/3/10	ESD-2	13:39:28.64	13:41:09.18	12	12	0	100,00%
6	19/4/10	ESD-3T	19:21:52.85	19:23:21.77	86	81	5	94,19%
7	20/4/10	ESD-3T	0:07:30.38	00:09:13.24	41	37	4	90,24%
8	19/5/10	ESD-3T	10:57:40.56	10:59:51.36	46	38	8	82,61%
9	18/6/10	ESD-3T	20:16:00.19	20:21:20.20	113	108	5	95,58%

Fig. 4. Results. Alarms suppressed

5   Conclusions

In this paper, we have presented a alarm management system that provides a solution for improving operators' situation awareness during emergency situations in offshore oil platforms. Oil process plant is a complex artifact composed of independent subparts that interacts with each other. The results of initial experiments run in our research lab using actual data information coming from SD scenarios have shown that only 6 percent of the total of the alarms were visualized to the operator which is an outstanding result. Additionally, operators confirmed that the suppressed information was unnecessary.

References

1. Health and Safety Executive, The explosion and fires at the Texaco Refinery, Milford Haven July 24, 1994: A report of the investigation by the Health and Safety Executive into the explosion and fires on the Pembroke Cracking Company Plant at the Texaco Refinery, Milford Haven on July 24, 1994 (1997), ISBN 071761413 1
2. Principais Acidentes da Indústria Petrolífera no Mundo,
   `http://www.ambientebrasil.com.br/composer.php3?base=./agua/salgada/`
   `index.html&conteudo=./agua/salgada/vazamentos.html`
3. Ajudante cai dentro de silo de soja e morre asfixiado, Brasil (August 31, 2002),
   `http://www.sindicatomercosul.com.br/noticia02.asp?noticia=5178`
4. Frente Nacional dos Petroleiros, Histórico dos Acidentes e Mortes na Petrobras–02 de outubro de (2008), `http://www.sindipetroalse.org.br/site/images/`
   `stories/Saude/histnapetrobras.pdf`
5. Murez, J., Berwanger, P.C.: Apparatus and method for performing process hazard analysis. US Patent 7, 716, 239 (2010)
6. Asea Brown Boveri (ABB), `http://www.abb.no/oilandgas`
7. GE Intelligent Platforms, `http://www.automation.com/pdf_articles/ge/alarm_`
   `response_management_wp_gfa789.pdf`
8. Rabuzin, K., Maleković, M., Baca, M.: A Combination of Reactive and Deliberative agents in Hospital Logistics. In: The Proceedings of 17th International Conference on Information and Intelligent Systems, Vara'Min, Croatia, pp. 63–70 (2006)
9. Rabuzin, K., Maleković, M., Cubrilo, M.: Resolving Physical Conflicts in Multiagent Systems. In: The Third International Multi-Conference on Computing in the Global Information Technology, ICCGI 2008, pp. 193–199. IEEE (2008)
10. Rabuzin, K., Maleković, M.: Efficient Trigger Management in Multiagent Systems. In: Central European Conference on Information and Intelligent Systems (2008)

Adaptive Pitch Control for Robot Thereminist Using Unscented Kalman Filter

Takeshi Mizumoto, Toru Takahashi, Tetsuya Ogata, and Hiroshi G. Okuno

Graduate School of Informatics, Kyoto University, Japan
mizumoto@kuis.kyoto-u.ac.jp

Abstract. We present an adaptive pitch control method for a theremin playing robot in ensemble. The problem of the theremin playing is its sensitivity to the environment. This degrades the pitch accuracy because its pitch characteristics are time varying caused by, such as a co-player motion during the ensemble. We solve this problem using a state space model of this characteristics and an unscented Kalman filter. Experimental results show that our method reduces the pitch error the EKF and block-wise update method by 90% and 77% on average, and the robot can play a musical score of 72.9 cent error on average.

1 Introduction

Robots that play instruments with humans in ensembles are expected to facilitate an intuitive and natural human-robot interaction. Since such an interaction through music is possible without common linguistic knowledge, it can provide a common base beyond cultural barriers such as language or generation. From the point of view of entertainment robotics, such robots will provide an interactive entertainment in which people can participate. In existing solo music robots [8, 7, 5] people are passive audience, not participants. Recently, some human-robot ensembles are reported, such as a duo [6], a trio [4], and a quartet [9].

We have developed a human-robot ensemble [4] with Robot Thereminist [5], A theremin is an electronic instrument played only by the player's hand motion. Its sound is determined by the distances between the both hands and two antennae of a theremin, respectively. Because of this mechanism, the theremin has an advantage that it can be played with no physical contacts and disadvantage that it is sensitive to the *environmental capacitance* [5].

An adaptive control problem in dynamic environment is not limited to the theremin, e.g., the temperature or wetness of the instrument, or the tension of a guitar, although the conventional music robots assumed a static environment. Indeed, we empirically found that the environmental capacitance actually changes through time. Therefore, existing thereminist robot studies are inappropriate for such a dynamic environment; A lookup-table-based method [1] takes a long time to rebuilding. The two-phase playing method using a nonlinear model [5] assume the environment is stable after the calibration. Although [11] achieves iterative update using a linear pitch model in log-scale, the pitch control accuracy is less than [5].

We present an adaptive pitch control using an unscented Kalman filter (UKF) [2] for dynamic environment. the UKF has three advantages than the extended Kalman filter

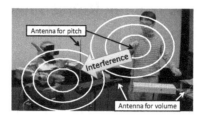

Fig. 1. The theremin played by a robot (left) is affected by the co-player(right)

(EKF) [10], (1) it can work with nonlinear functions since it needs no Jacobians. (2) It estimates the hidden states by the second order, (3) The sigma points, i.e., the samples for state estimation, are selected deterministically. Thus, the number of samples is lower than probabilistic methods, e.g., particle filters [3],

2   Static Pitch Control for Robot Thereminist

We describe a parametric model of the pitch characteristics [5]. Let p be the pitch, $\theta = (\theta_0, \theta_1, \theta_2, \theta_3)$ be the model parameters of the pitch model, and $x_p \in [0, 1]$ be the normalized robot's arm position. $x_p = 1$ and $x_p = 0$ means the closest and farthest arm position, respectively. The pitch characteristics M_p, i.e., the pitch p when the arm is at x_p, are experimentally formulated as:

$$p = M_p(x_p; \theta) = \theta_2/(\theta_0 - x_p)^{\theta_1} + \theta_3, \text{and} \tag{1}$$

$$x_p = M_p^{-1}(p; \theta) = \theta_0 - (\theta_2/(p - \theta_3))^{1/\theta_1} \tag{2}$$

The static pitch control method consists of *calibration* and *performance* phase [5]. In the calibration phase, the model parameter $\hat{\theta}$ is estimated using the $L + 1$ pairs of the arm position $x_p^{(i)}$ and the observed theremin's pitch $p^{(i)}$ at the position ($i = 0, ..., L$): $x_p^{(i)} = i/L$, and $p^{(i)} = M_p(x_p^{(i)}; \theta_T)$ where θ_T is the unknown true parameter. Then, $\hat{\theta}$ is estimated by minimizing the square error, $\sum_{i=0}^{L} ||p^{(i)} - M_p(x_p^{(i)}; \theta)||^2$, using the Levenberg-Marquardt method. In the performance phase, the robot's arm position is calculated from a given musical score, defined as the desired pitch trajectory $q(t)$. The arm position trajectory $\hat{x}_p(t)$ is defined with $\hat{\theta}$ and $q(t)$ as $\hat{x}_p(t) = M_p^{-1}(q(t); \hat{\theta})$.

The conventional methods [5,1] assume that θ_T is time invariant after the calibration. In ensemble, however, θ_T is time varying because of the gradual change, such as room temperature or a theremin's internal state, and the co-player's motion.

3   Adaptive Pitch Control Using Unscented Kalman Filter

The problem statement of the adaptive pitch control is:

Inputs   Desired pitch trajectory $q(t)$ (i.e., a musical score),
               initial parameter $\theta(0)$, and observed pitch $p(t)$
Outputs   Estimated parameter $\hat{\theta}(t)$ and arm position $\hat{x}_p(t) = M_p^{-1}(q(t); \hat{\theta}(t))$
Assumption   The true parameter $\theta_T(t)$ performs a random walk.

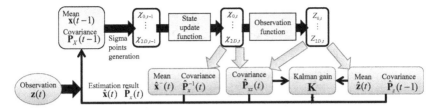

Fig. 2. Data flow of unscented Kalman filter

At time t, the robot moves its arm to $\hat{x}_p(t)$ using $q(t)$ and $\hat{\theta}(t)$, then, observes the played pitch $p(t)$. $p(t)$ and $q(t)$ are different because $p(t)$ is generated from the true parameter $\hat{\theta}_T(t)$. The objective of this problem is to minimize $|p(t) - q(t)|^2$ by updating the estimation of $\hat{\theta}(t)$. We have two candidates: block-wise update and Kalman filtering. The former accumulates observations and re-estimates the $\hat{\theta}$, i.e., a periodic calibration. The latter updates the parameters for each observation.

We solve the problem using UKF. It estimates a hidden state using an unscented transform [2], which estimates the statistics after any nonlinear transforms using deterministically selected samples. The adaptive pitch control is the same method as in section 2 since the $\theta(t)$ is updated through time. Figure 2 is overview of the UKF. See [2] for complete description.

We build a state space model to utilize the UKF. Let the state be the model parameter $\theta \in R^{1 \times 4}$. The state update is a random walk, and the observation is M_p in Eq. (1). The state space model is summarized as follows.

$$\text{State update}: \quad \theta(t) = f(\theta(t-1), \mathbf{v}(t-1)) = \theta(t-1) + \mathbf{v}(t-1) \tag{3}$$

$$\theta_0(t) = \max(\theta_0(t), 1 + \varepsilon) \tag{4}$$

$$\theta_1(t) = \max(\theta_1(t), 0) \tag{5}$$

$$\text{Observation}: \quad p(t) = h(\theta(t), w(t)) = M_p(x_p(t); \theta(t)) + w(t) \tag{6}$$

where $\mathbf{v}(t) \in R^{1 \times 4}$ and $w(t) \in R$ denote the state update noise and the observation noise, respectively. Eqs. (4) and (5) are additional constraints to ensure that the M_p in Eq. (1) is real and finite. Although these constraints make indifferentiable points, the state-update function is still valid for the UKF because it requires no Jacobian.

If we simply substitute Eq. (2) with Eq. (6), the function is $p(t) = q(t)$, which is not a mapping of the θ; i.e., Eq. (6) does not *observe*. This is not true because the parameters used for Eqs. (1) and (2) are different. Let θ_T be the true parameter of the environment. Then, the observation function $p(t)$ after substitution is a nonlinear function $p(t) = M_p(M_p^{-1}(q(t); \hat{\theta}); \theta_T)$. If $\hat{\theta} \neq \theta_T$, i.e., estimation is imperfect, M_p and M_p^{-1} are not the inverse functions. Therefore, Eq. (6) is a nonlinear mapping until the parameter estimation succeeds. The parameter estimation is unlikely to become perfect because the unknown noise $\mathbf{v}$ fluctuate the true parameter θ_T.

Considering the limitation of the robot's arm speed, we add a constraint using the maximum arm speed x_{plim}: $x_{plim} > |x_p(t) - x_p(t+1)|$. If it is not satisfied, the $p_t(t+1)$ is $x_p(t+1) = x_p(t) + x_{plim}$. We empirically set $x_{plim} = 0.05$ in the experiments.

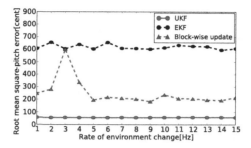

Fig. 3. Pitch errors in simulation: the vertical and horizontal axes denote the pitch error and the environment change speed, i.e., ω of $e(t)$. A red solid, black dashed, and blue dashed lines denote the UKF, the EKF, and the block-wise update results, respectively.

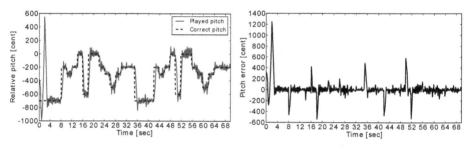

Fig. 4. The black broken and the red lines denote the desired and played pitch trajectories

Fig. 5. The horizontal axis denotes the time, and the vertical axis denotes the pitch error

4   Experiments

4.1   Evaluation Using Simulation

To simulate a realistic environmental dynamics, we generate a model parameters from N measurements; $\theta^{(0)},..,\theta^{(N)}$. We construct a function $\theta = f(e)$ by linearly interpolating the pairs of $(e,\theta) = (i/N, \theta^{(i)})$. A time series of $\theta_T(t)$ is determined by designing the time series of $e(t)$. We define $e(t) = 0.5\sin(2\pi\omega) + 0.5$, $\omega = 1,...,15$, to simulate a situation that co-player is going closer to and away from the theremin.

We compare three methods; the UKF, the EKF, and block-wise update. The parameters for UKF are $\Sigma_x = \Sigma_v = diag(5,5,5,5)$, $\Sigma_w = 10$, $\kappa = 2$, and $(D_x, D_v, D_z, D_w) = (4,4,1,1)$. Those for EKF, Σ_v and Σ_w, are the same as the UKF. We add a constraint $\theta_0 \geq x_p$ by substituting θ_0 with x_p to ensure the $\hat{\theta}(t)$ is finite. For the block-wise parameter estimation, the $\hat{\theta}$ is re-estimated for each 5 sec.

The evaluation criteria is a root-mean-square of the pitch error c [cent] defined by $c = 1200\log(p/q)$, where p and q denote the observed and desired pitch in Hz, respectively. If the pitch error is 100 [cent], it is equivalent to the half-note error. For each condition, we performed the experiment 10 times and averaged the results.

The result is summarized in Fig. 3. The UKF outperforms the others, because of its precise approximation. The EKF is the worst because of its poor approximation of observation function. Even if we set a constant to $e(t)$, it fails because of the error accumulation. The block-wise update has the middle error of UKF and EKF because it achieves a precise but slow estimation. It has an especially high error at $\omega = 3$. This suggests the sensitivity to the timing of the re-estimation. In the worst case, the environment changes just after the parameter re-estimation, e.g., the $e(t)$ for parameter estimation is $e(t) = 1$, whereas that for the theremin playing is $e(t) = 0$.

4.2   Evaluation Using a Robot

We implement our method on a humanoid robot, HRP-2, to demonstrate the performance on a real robot. The robot's arm position and the model parameters are updated for each 62.5 msec time interval. The pitch is estimated with a auto-correlation-based method [5]. The initial parameter is set to be a wrong one to demonstrate its adaptation. An American folk song "Aura Lee" is used as a score.

Fig. 4 demonstrates that the robot's play. A pitch is calculated with a relative pitch from 220 Hz in cent. As shown in the figure, the robot plays the pitch correctly totally. The pitch is wrong from 0 to 4 sec because the estimated parameters had not converged yet. After the convergence, the succeeding pitches are correct. There remain a fluctuation after the convergence because (1) the interval difference of pitch estimation and arm position measurement, and (2) the quick pitch change makes physical oscillation. Fig. 5 shows the pitch error of Fig. 4. The mean absolute error is 72.9 [cent], i.e., less than the half-note error. The peaks in trajectory at 8, 16 and 36 seconds occurred because of the convergence time

5   Conclusion

We presented an adaptive pitch control method for thereminist robot in a dynamic environment. We developed a state space model of a time varying pitch characteristics, and tracking method using UKF. Experimental results showed that our method outperforms than both EKF and a block-wise update method, and demonstrated on a robot. For the next step, we will implement a human-robot ensemble system.

Acknowledgements. This study was partially supported by a Grant-in-Aid for Scientific Research (S) (No. 19100003), a Grant-in-Aid for Scientific Research on Innovative Areas (No. 22118502), and the Global COE Program.

References

1. Alford, A., Northrup, S., Kawamura, K., Chan, K.W., Barile, J.: A music playing robot. In: Proc. FSR, pp. 29–31 (1999)
2. Julier, S.J., Uhlmann, J.K.: Unscented filtering and nonlinear estimation. Proc. IEEE 92(3), 401–422 (2004)
3. Kitagawa, G.: Monte carlo filter and smoother for non-gaussian nonlinear state space models. J. of Computational and Graphical Statistics 5(1), 1–25 (1996)

4. Mizumoto, T., Lim, A., Otsuka, T., Nakadai, K., Takahashi, T., Ogata, T., Okuno, H.G.: Integration of flutist gesture recognition and beat tracking for human-robot ensemble. In: Proc. IROS, Robots and Musical Expressions, pp. 51–56 (2010)
5. Mizumoto, T., Tsujino, H., Takahashi, T., Ogata, T., Okuno, H.G.: Thereminist robot: Development of a robot theremin player with feedforward and feedback arm control based on a theremin's pitch model. In: Proc. IROS, pp. 2297–2302 (2009)
6. Petersen, K., Solis, J., Takanishi, A.: Development of a aural real-time rhythmical and harmonic tracking to enable the musical interaction with the waseda flutist robot. In: Proc. IROS, pp. 2303–2308 (2009)
7. Solis, J., Petersen, K., Ninomiya, T., Takeuchi, M., Takanishi, A.: Development of anthropomorphic musical performance robots: From understanding the nature of music performance to its application to entertainment robotics. In: Proc. IROS, pp. 2309–2314 (2009)
8. Sugano, S., Kato, I.: WABOT-2: Autonomous robot with dexterous finger-arm - finger-arm coordination control in keyboard performance. In: Proc. ICRA, pp. 90–97 (1987)
9. Weinberg, G., Blosser, B., Mallikarjuna, T., Raman, A.: The creation of a multi-human, multi-robot interactive jam session. In: Proc. NIME, pp. 70–73 (2009)
10. Welch, G., Bishop, G.: An introduction to the kalman filter. In: Proc. SIGGRAPH Course, vol. (8) (2001)
11. Wu, Y., Kuvinichkul, P., Cheung, P., Demiri, Y.: Towards anthropomorphic robot thereminist. In: Proc. ROBIO, pp. 235–240 (2010)

Cognitive Computing and Affective Computing

An Adaptive Agent Model for the Emergence
of Recurring Dream Scripts Based on Hebbian Learning

Valérie de Kemp and Jan Treur

VU University Amsterdam, Agent Systems Research Group
De Boelelaan 1081, 1081 HV, Amsterdam, The Netherlands
v.p.de.kemp@vu.nl, treur@cs.vu.nl
http://www.cs.vu.nl/~treur

Abstract. In this paper an adaptive agent model is presented that models how recurrent dreams such as nightmares occur. In the agent model recurring dreams emerge as scripts of connected dream episodes in which the connections between the episodes strengthen over time by Hebbian learning. The model was evaluated by a number of simulation experiments.

1   Introduction

In the recent cognitive and neurological literature the mechanisms and functions of dreaming have received much attention; e.g., [19-23], [28-32]. Dreaming makes use of memory elements for sensory representations (mental images) and their associated emotions to generate 'virtual simulations'; e.g., [20], pp. 499-500. Usually dreaming is considered a form of internal simulation of real-life-like processes serving as training in order to learn or adapt certain capabilities. Since a long time these virtual stories and their possible interpretations have been used in a therapeutical context; e.g., [8], [9]. In [26] and [27] computational models for some aspects of dreaming are contributed. The models presented in these and the current paper may be a basis for formalisation of this area and training of therapists.

The adaptive agent model presented here adopts basic elements from [26]. However, in contrast to this, the agent model introduced here generates coherent scripts (virtual stories) for the type of internal simulation that is assumed to take place in dreaming. The model presented here does not consider dreams as consisting of isolated episodes as in [26], but models them as coherent scripts with mutually connected mental images and associated feelings to represent the virtual story line, following what is pointed out in [25]. Moreover, the connections between them are strengthened by Hebbian learning, thus developing a recurrent pattern as in recurring nightmares; e.g., [21], [25]. Similar to [26] it incorporates emotion regulation to suppress too high feeling levels and sensory representation states.

In the paper in Section 2 some background literature and the adaptive agent model is described in more detail. Section 3 presents simulation results providing some dream scenarios. Finally, Section 4 is a discussion, in which also the relation of the model with neurological theories and findings is addressed.

W. Ding et al. (Eds.): Modern Advances in Intelligent Systems and Tools, SCI 431, pp. 27–35.
springerlink.com   © Springer-Verlag Berlin Heidelberg 2012

2   The Adaptive Agent Model for Recurring Dream Scripts

In this section, first it is discussed how in dreaming memory elements with their associated emotions are used as building blocks for an internal simulation of real life. Furthermore, it is pointed out how such internal simulations can lead to dream scripts consisting of fixed sequences of episodes that become recurrent dreams. After this the formalised agent model incorporating these elements is presented.

Within the literature the role of memory elements providing content for dreams is well-recognized; e.g., [21], p. 499-500. In particular, it is recognized that the choice for memory elements with some emotional association and (re)combining them into a dream facilitates emotion generation: the emotional associations of the sensory memory elements may make that a person has to cope with high levels of emotions (e.g., fear) felt in the dream; cf. [21], p. 500. *Emotion regulation* mechanisms are used to suppress emotions that are felt as too strong; cf. [11], [13], [14]. Thus dreams are considered as flows of activated sequences of images based on (re)combined memory elements and associated emotions; cf. [21], p. 500. Such flows can be related to the notion of *internal simulation* put forward, among others, by [6], [7], [12], [17], [18].

The idea of internal simulation is that sensory representation states are activated (e.g., mental images), which in response trigger associated preparation states for actions or bodily changes, which, by prediction links, in turn activate other sensory representation states. The latter states represent the effects of the prepared actions or bodily changes, without actually having executed them. Being inherently cyclic, the simulation process can go on indefinitely. Internal simulation has been used, for example, to describe (imagined) processes in the external world (e.g., prediction of effects of own actions [3]), or processes in another person's mind (e.g., emotion recognition or mindreading [12]) or processes in a person's own body (e.g., [6]). Although usually internal simulation as briefly described above concerns mental processes for awake persons, it is assumed that it may be applicable as well to describe dreaming.

The idea of internal simulation has been exploited in particular by applying it to bodily changes relating to emotions, using the notion of *as-if body loop* [6]. To describe the generation of emotions and feelings, in [23] a causal chain from sensory representation to preparation of emotional response to feeling the bodily effect of the emotional response was introduced: an *as-if body loop* (cf. [6], pp. 155-158; [7], pp. 79-80; [9]). An as-if body loop describes an inner simulation of bodily processes, without actually affecting the body. Note that in [6] an emotion (or emotional response) is distinguished from a feeling (or felt emotion). The emotion and feeling mutually affect each other: an as-if body loop usually occurs in a cyclic form by assuming that the emotion felt in turn also affects the prepared bodily changes, as he points out, for example, in ([8], pp. 91-92; [9], pp. 119-122).

One theory explicitly referring to a purpose of dreaming as internal simulation is the threat simulation theory of the evolutionary function of dreaming (cf. [23], [28]). This theory assumes that dreaming is an evolutionary adaptation to be able to rehearse coping with threatening situations in a safe manner. Others consider the function of dreaming in strengthening the emotion regulation capabilities for fear; e.g., [15], [29],

[30]. For both purposes adequate exercising material is needed for the dreams: fearful situations have to be imagined, built on memory elements suitable for fear arousal.

More specific inspiration for the adaptive agent model for the development of recurrent dreams introduced here was taken from the informal description in [25]. This describes how a nightmare is stored in memory as a *script*: a scary series of events. The probability that the (nightmare) script is activated during the REM sleep depends on the accessibility of the script. If the script is associated with much fear, the accessibility is high. Initially a dream may be neutral, but if there is some association to the nightmare script, the script may be activated. To activate associations to the script, a form of resemblance interpretation plays a role. If the dreamer is anxious, neutral dream elements tend to be interpreted as threatening. If an ambiguous stimulus is perceived as threatening it will be perceived as more similar to a threatening script than if it is perceived as neutral. The replaying of the nightmare over and over again causes nightmare distress. The more neurotic (biased towards the negative side) the dreamer is, the more the dreamer gets distressed. The more the dreamer is distressed, the better the script is consolidated, the more likely it is that the nightmare will become recurrent.

An overview of the agent model is shown in Fig. 1. For an explanation of the symbols used, see Table 1. It reuses some parts of the model described in, [26]. The neuroticism state ns is added, which indicates how neurotic the dreamer is. The more neurotic the dreamer is, the more he or she is emotionally affected by the dream, which leads to a higher value for the feeling state fs_b. Moreover, connections between the sensory representations srs_{sk} for different stimuli were added, which makes it possible to activate related stimuli to the ones already active. With strong connections between the sensory representations, the same kind of dream can be replayed, making it possible to have a recurrent nightmare. The agent model has been formalised as a set of differential equations. During processing, each state has an activation level represented by a real number between 0 and 1. Below, the (temporally) Local Properties (LP) for the dynamics based on the connections between the states in Fig. 1 are described.

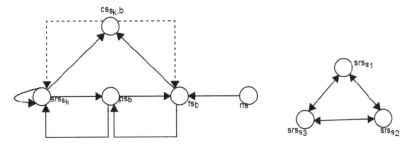

Fig. 1. Elements of the agent model for the generation of recurrent dreams

Table 1. Explanation of the states used in the model

state	explanation
ps_b	Preparation state for bodily response b
fs_b	Feeling state for b
srs_{s_k}	Sensory representation state for s_k
$cs_{s_k,b}$	Control state for regulation of sensory representation of s_k and feeling b
es_{s_k}	Episode state for s_k
mt_{s_k}	Memory trigger for s_k
ns	Neuroticism state

In the specifications below a threshold function *th* is used as a combination function for k incoming connections as follows: the combined input level is $th(\mu_1 V_1 + \ldots + \mu_k V_k)$ with μ_i the connection strength for incoming connection i and V_i the activation level of the corresponding connected state. In LP1 to LP4 the following continuous logistic form was used:

$$th(X) = (1/(1+e^{-\sigma(X-\tau)}) - 1/(1+e^{\sigma\tau}))\,(1+e^{-\sigma\tau})$$

Here σ is a steepness and τ a threshold parameter. Note that for higher values of $\sigma\tau$ (e.g., σ higher than $20/\tau$) this threshold function can be approximated by the simpler expression; this has been used in LP5: $th(X) = 1/(1+e^{-\sigma(X-\tau)})$. Table 2 shows the connection strengths used in the specification of the agent model described below. The first property LP1 describes how preparation for response b is affected by the sensory representation and episode states of stimuli s_k (triggering the response), and by the feeling state for b. Parameter γ indicates the speed by which an activation level is updated upon received input from other states.

Table 2. Connections and weights used in the model

from states	to state	weights	LP
$srs_{s_1}, \ldots, srs_{s_n}, fs_b$	ps_b	$\omega_{11}, \ldots, \omega_{1n}, \omega_2$	LP1
$ps_b, cs_{s_1, b}, \ldots, cs_{s_n, b}, ns$	fs_b	$\omega_3, \omega_{41}, \ldots, \omega_{4n}, \omega_{12, 0}$	LP2
$ps_b, cs_{s_k, b}, mt_{s_k}, srs_{s_i}$	srs_{s_k}	$\omega_{5k}, \omega_{6k}, \omega_{0k}, \omega_{14, ik}$	LP3
srs_{s_k}, fs_b	$cs_{s_k, b}$	ω_{7k}, ω_{8k}	LP4
$srs_{s_k}, es_{s_1}, \ldots, es_{s_n}, cs_{s_k, b}$	es_{s_k}	$\omega_{9k}, \omega_{10,1k}, \ldots, \omega_{10, nk}, \omega_{11, k}$	LP5

LP1 Preparation state for emotional response b
If   sensory representation states of s_k ($k = 1, 2, \ldots$) have level V_{1k}
and   the feeling state for b has level V_2 and the preparation for b has level V_3
then   after Δt the preparation state for b will have
            level $V_3 + \gamma\,[th(\Sigma_k\omega_{1k}V_{1k} + \omega_2 V_2) - V_3]\,\Delta t$.
    $d\,ps_b(t)/dt = \gamma\,[th(\Sigma_k\omega_{1k}srs_{s_k}(t) + \omega_2\,fs_b(t)) - ps_b(t)]$

LP2 Feeling state for b
If   the preparation state for b has level V_1
and   the control states for s_k and b ($k=1, \ldots, n$) have levels V_{2k}
and   the neuroticism state has level V_3

and   the feeling state for b has level V_4
then   after Δt the feeling state for b will have
        level $V_4 + \gamma\, [\ th(\omega_3 V_1 + \Sigma_k\, \omega_{4k} V_{2k} + \omega_{12,0} V_3) - V_4]\ \Delta t.$
        $d\, fs_b(t)/dt = \gamma\, [\ th(\omega_3\, ps_b(t) + \Sigma_k\, \omega_{4k}\, cs_{sk,\, b}(t) + \omega_{12,0}\, ns(t)) - fs_b(t)]$

LP3   Sensory representation state for s_k
If   the preparation state for b has level V_1
and   the control state for s_k and b has level V_{2k}
and   the memory trigger for s_k has level V_{3k}
and   the sensory representation states of s_i ($i=1, ..., n$) have level V_{4i}
and   the sensory representation state for s_k has level V_{5k}
then   after Δt the sensory representation state for sk will have
        level $V_{6k} + \gamma\, [\ th(\omega_{5k} V_1 + \omega_{6k}\, V_{2k} + \omega_{0k} V_{3k} + \Sigma_i\, \omega_{14,\, ik}\, V_{4i}) - V_{5k}]\ \Delta t.$
        $d\, srs_{sk}(t)/dt = \gamma\, [\ th(\omega_{5k}\, ps_b(t) + \omega_{6k}\, cs_{sk,\, b}(t) + \omega_{0k}\, mt_{sk}(t) + \Sigma_i\, \omega_{14,\, ik}\, srs_{si}(t)) - srs_{sk}(t)]$

LP4   Control state for s_k and b
If   the sensory representation state for s_k has level V_{1k}
and   the feeling state for b has level V_2
and   the control state for s_k and b has level V_{3k}
then   after Δt the control state for s_k and b will have
        level $V_{3k} + \gamma\, [\ th(\omega_{7k} V_{1k} + \omega_{8k} V_2) - V_{3k}]\ \Delta t.$
        $d\, cs_{sk,\, b}(t)/dt = \gamma\, [\ th(\omega_{7k}\, srs_{sk}(t) + \omega_{8k}\, fs_b(t)) - cs_{sk,\, b}(t)]$

At each point in time multiple sensory representation states can be active simultaneously. For cases of awake functioning the *Global Workspace Theory* (cf. [1]) was developed to describe how a single flow of conscious experience can come out of such a large multiplicity of (unconscious) processes. The basic idea is that a *winner-takes-it-all competition* takes place to determine which one will get dominance and be included in the single flow of consciousness (after which it is accessible to all processes). This idea was applied here in the dreaming context to determine which sensory representation element will be included as an episode state es_{sk} in a dream episode. This competition process is decribed in LP5, using inhibiting connections from the episode states es_{si} with $i \neq k$ to es_{sk}. For the suppressing effects the connection weights from the es_{si} with $i \neq k$ to es_{sk} are taken negative. Note that for the sake of notational simplicity $\omega_{10,kk} = 0$ is taken. For traumatic stimuli s_k an additional and strong way of inhibition of the corresponding episode state takes place, blocking the generation of an episode state for this stimulus. It is based on the control state for s_k and b and is assumed to have a strong negative connection strength ω_{e3k}. For non-traumatic stimuli this connection strength is 0.

LP5   Episode state for s_k
If   the sensory representation state for s_k has level V_{1k}
and   the control state for s_k and b has level V_{2k}
and   the episodic states for s_i ($i = 1, ...$) have level V_{3i}
then   after Δt the episodic state for s_k will have
        level $V_{3k} + \gamma\, [th(\omega_{9k} V_{1k} + \omega_{11,k} V_{2k} + \Sigma_i\, \omega_{10,ik} V_{3i}) - V_{3k}]\ \Delta t.$
        $d\, es_{sk}(t)/dt = \gamma\, [\ th(\omega_{9k}\, srs_{sk}(t) + \omega_{11,k}\, cs_{sk,\, b}(t) + \Sigma_i\, \omega_{10,ik}\, es_{si}(t)) - es_{sk}(t)]$

Hebbian Learning for the Connections between Sensory Representations

From a Hebbian perspective [16], strengthening of a connection over time may take place when both nodes are often active simultaneously ('neurons that fire together wire together'). The principle goes back to Hebb [16], but has recently gained enhanced interest by more extensive empirical support (e.g., [2]), and more advanced mathematical formulations (e.g., [10]). In the adaptive agent model the mutual connections between the sensory representations for s_k are adapted based on a Hebbian learning principle. More specifically, for such a connection from node i to node j its strength ω_{ij} is adapted using the following *Hebbian learning rule*, taking into account a maximal connection strength *1*, a *learning rate* η, and an *extinction rate* ζ (usually taken small):

$$d\omega_{ij}(t)/dt = \eta \, a_i(t)a_j(t)(1 - \omega_{ij}(t)) - \zeta\omega_{ij}(t) = \eta \, a_i(t)a_j(t) - (\eta \, a_i(t)a_j(t) + \zeta) \, \omega_{ij}(t)$$

Here $a_i(t)$ and $a_j(t)$ are the activation levels of node i and j at time t and $\omega_{ij}(t)$ is the strength of the connection from node i to node j at time t. A similar Hebbian learning rule can be found in [10], p. 406. By the factor $1 - \omega_{ij}(t)$ the learning rule keeps the level of $\omega_{ij}(t)$ bounded by 1 (which could be replaced by any other positive number). When the extinction rate is relatively low, the upward changes during learning are proportional to both $a_i(t)$ and $a_j(t)$ and maximal learning takes place when both are 1. Whenever one of $a_i(t)$ and $a_j(t)$ is 0 (or close to 0) extinction takes over, and ω_{ij} slowly decreases (unlearning).

3   Simulation Results

The agent model has been used to conduct a number of simulation experiments. These experiments are discussed more extensively in the Appendix at URL http://www.few.vu.nl/~wai/IEA12dreamscripts. In the simulations four stimuli play a role. To make the results more readable the following descriptions are added, which was inspired by [25].

> s_1: The dreamer is chased by un unknown person.
>
> s_2: The dreamer walks in an dark environment and sees an unknown person.
>
> s_3: The unknown person is lost and asks the dreamer the way.
>
> s_4: The dreamer can't get away quick enough and the chaser is getting near.

The following type of scenario have been realized in simulations:

- A memory trigger activates the sensory representation state of an emotionally neutral stimulus s_2. The corresponding episode state is triggered and the dream begins.
- Another stimulus, s_1, is associated with fear. There is a strong connection from the sensory representation state of s_2 to the sensory representation state of s_1. This is because the stimulus s_2 evokes an association with s_1, due to resemblance or feelings of fear. Therefore the sensory representation state of s_1 is also triggered.
- The stimulus s_3 is a reassuring stimulus, which does not go together with stimuli which evoke feelings of fear like s_1. The stimuli s_1 and s_3 contradict each other. The sensory representation state of s_2 does also have an excitatory influence on the

sensory representation state of s_3. It means that when the sensory representation state of s_2 is activated the sensory representation states of both s_1 and s_3 are both triggered. Only one of the episodes of s_1 and s_3 is activated above threshold at the same time.

- The stimulus s_1 is associated with fear, which means that if the sensory representation state of s_1 is active, the preparation state for bodily response b and the feeling state for b have higher levels than usual. The more neurotic the dreamer is the higher the level of the feeling state for b will be. The higher the activation of the feeling state for b, the more it inhibits the sensory representation state of s_3 and the more it excites the sensory representation states of s_3 and s_4.
- The stimulus s_4 is associated with more fear than s_1. When the level of the feeling state for b is getting higher the level of the sensory representation state for s_4 catches up with the level of the sensory representation state for s_1 and the episode of s_1 is followed by the episode of s_4. The nightmare becomes really frightening.

4   Discussion

Dreaming generates 'virtual simulations' based on memory elements for sensory representations (mental images) and their associated emotions; e.g., [20]. Although in the recent cognitive and neurological literature the mechanisms and functions of dreaming have received much attention (e.g., [19-23], [28-32]), only very few computational models are known that address aspects of dreaming; for example, see [26] and [27]. Sometimes dreams emerge that get a recurring character, for example, nightmares; cf. [21], [25]. The presented adaptive agent model shows how recurrent dreams can emerge. The agent model uses adapting connections between different sensory representations and by a converging process obtains a fixed script of dream episodes that can be activated time and time again. The adaptation mechanism is based on Hebbian learning. The model builds further on a previous model for dreaming described in [26]. By this existing model the adaptive scripting process which is the basis of the emergence of recurring dreams and nightmares is not covered. It has been shown that by the adaptive agent model introduced here indeed recurring dreams following a fixed script emerge. In [27] a computational model is described that focuses on fear extinction learning during dreaming. In future work an extension of the agent model presented here may be made which also incorporates fear extinction mechanisms as in [27].

References

1. Baars, B.J.: In the theater of consciousness: the workspace of the mind. Oxford University Press, Oxford (1997)
2. Bi, G., Poo, M.: Synaptic modification by correlated activity: Hebb's postulate revisited. Annu. Rev. Neurosci. 24, 139–166 (2001)
3. Becker, W., Fuchs, A.F.: Prediction in the Oculomotor System: Smooth Pursuit During Transient Disappearance of a Visual Target. Exp. Brain Research 57, 562–575 (1985)

4. Damasio, A.R.: Descartes' Error: Emotion, Reason and the Human Brain. Papermac, London (1994)
5. Damasio, A.R.: The Feeling of What Happens. Body and Emotion in the Making of Consciousness. Harcourt Brace, New York (1999)
6. Damasio, A.R.: Looking for Spinoza: Joy, Sorrow, and the Feeling Brain. Vintage books, London (2003)
7. Damasio, A.R.: Self comes to mind: constructing the conscious brain. Pantheon Books, NY (2010)
8. Faraday, A.: Dream power. Coward, Mccann & Geoghegan, Oxford, England (1972)
9. Freud, S.: The Interpretation of Dreams (First German print 1900). Hogarth Press, London (1953)
10. Gerstner, W., Kistler, W.M.: Mathematical formulations of Hebbian learning. Biol. Cybern. 87, 404–415 (2002)
11. Goldin, P.R., McRae, K., Ramel, W., Gross, J.J.: The neural bases of emotion regulation: reappraisal and suppression of negative emotion. Biol. Psychiatry 63, 577–586 (2008)
12. Goldman, A.I.: Simulating Minds: The Philosophy, Psychology, and Neuroscience of Mindreading. Oxford Univ. Press, New York (2006)
13. Gross, J.J.: Antecedent- and response-focused emotion regulation: divergent consequences for experience, expression, and physiology. J. of Personality and Social Psych. 74, 224–237 (1998)
14. Gross, J.J.: Handbook of Emotion Regulation. Guilford Press, New York (2007)
15. Gujar, N., McDonald, S.A., Nishida, M., Walker, M.P.: A Role for REM Sleep in Recalibrating the Sensitivity of the Human Brain to Specific Emotions. Cerebral Cortex 21, 115–123 (2011)
16. Hebb, D.: The Organisation of Behavior. Wiley (1949)
17. Hesslow, G.: Will neuroscience explain consciousness? J. Theoret. Biol. 171, 29–39 (1994)
18. Hesslow, G.: Conscious thought as simulation of behaviour and perception. Trends Cogn. Sci. 6, 242–247 (2002)
19. Hobson, J.A.: REM sleep and dreaming: towards a theory of protoconsciousness. Nature Reviews Neuroscience 10, 803–814 (2009)
20. Levin, R., Nielsen, T.A.: Disturbed dreaming, posttraumatic stress disorder, and affect distress: A review and neurocognitive model. Psychological Bulletin 133, 482–528 (2007)
21. Levin, R., Nielsen, T.A.: Nightmares, bad dreams, and emotion dysregulation. A review and new neurocognitive model of dreaming. Curr. Dir. Psychol. Sci. 18, 84–88 (2009)
22. Nielsen, T.A., Stenstrom, P.: What are the memory sources of dreaming? Nature 437, 1286–1289 (2005)
23. Revonsuo, A.: The reinterpretation of dreams: An evolutionary hypothesis of function of dreaming. Behavioral and Brain Sciences 23, 877–901 (2000)
24. Salzman, C.D., Fusi, S.: Emotion, Cognition, and Mental State Representation in Amygdala and Prefrontal Cortex. Annu. Rev. Neurosci. 33, 173–202 (2010)
25. Spoormaker, V.I.: A cognitive model of recurrent nightmares. International Journal of Dream Research 1, 15–22 (2008)
26. Treur, J.: A Computational Agent Model Using Internal Simulation to Generate Emotional Dream Episodes. In: Samsonovich, A.V., Jóhannsdóttir, K.R. (eds.) Proceedings of the Second International Conference on Biologically Inspired Cognitive Architectures, BICA 2011, pp. 389–399. IOS Press (2011)

27. Treur, J.: Dreaming Your Fear Away: A Computational Model for Fear Extinction Learning During Dreaming. In: Lu, B.-L., Zhang, L., Kwok, J. (eds.) ICONIP 2011, Part III. LNCS, vol. 7064, pp. 197–209. Springer, Heidelberg (2011)
28. Valli, K., Revonsuo, A., Palkas, O., Ismail, K.H., Ali, K.J., Punamaki, R.L.: The threat simulation theory of the evolutionary function of dreaming: evidence from dreams of traumatized children. Conscious Cogn. 14, 188–218 (2005)
29. Walker, M.P.: The role of sleep in cognition and emotion. Ann. N. Y. Acad. Sci. 1156, 168–197 (2009)
30. Walker, M.P., van der Helm, E.: Overnight therapy? The role of sleep in emotional brain processing. Psychol. Bull. 135, 731–748 (2009)
31. Yoo, S.S., Gujar, N., Hu, P., Jolesz, F.A., Walker, M.P.: The human emotional brain without sleep – a prefrontal amygdala disconnect. Curr. Biol. 17, R877–R878 (2007)

Application of Things: A Step beyond Web of Things

J.A. Ortega Ramírez[1], L.M. Soria Morillo[1], Alexander Kröner[2],
and J.A. Álvarez García[1]

[1] University of Seville, Seville, Spain
{jortega,lsoria,jaalvarez}@us.es
[2] German Research Center for Artificial Intelligence, Saarbrucken, Germany
alexander.kroener@dfki.de

Abstract. Internet of Things has achieved a great expectation in the last few years, largely due to possibilities in the field of the interaction *between* user and environment. This interaction allows users to get involved in their environments, managing at all times each item that surrounds them. Specially, because of complex services delivered by surrounding item, increasingly it is necessary to help users to choose what services are offered and what are theirs functionalities. Is this paper, is presented a new approach to the human-items interaction by using the concept of Internet of Things. In this way, the user can not only run the different applications offers by smart-things, but to share theirs results through many social networks and allows the remote execution of services provided by the own user and his/her social environment. Furthermore, a structure will be defined to assure the best user comfort when information sharing between the device and the environments was needed.

Keywords: RFID tag, Information sharing, Web of Things, Mobile environments, Pervasive computing.

1 Introduction

1.1 Motivation

Since in 1999 the concept of Internet of Things born from the hands of Roy Want (W. Roy, 1999), many things have changed. First, the society has suffered a technological revolution and everyone interact in their daily lives with electronic devices like mobile phones, electronic agendas, personal computers and so on. The number of functionalities that these elements give to people is huge and sometimes, can be reached to overload the environments. The first sequel of this fact is that services potentials are not exploited by users and in most cases the functionalities used are the simplest ones. This problem occurs often with elderly people or with non-technological users.

Internet of Things can help to these users by providing a way to shows the available services that every smart-thing surrounding them can offer. Obviously, when we refer to smart-things not only make reference to electronic devices, but every object that can be identified by using the technology. In this field lies the potential of IoT, due to in most of related works are used RFID or NFC (Near Field Communication)

W. Ding et al. (Eds.): Modern Advances in Intelligent Systems and Tools, SCI 431, pp. 37–46.

tags to identify the objects (Kranz, Holleis, & Schmidt, 2010) (Alexander Kröner, 2008). RFID and NFC are technologies that allow writing in a magnetic tag some information so that later, that information will be able to be read by using many sensors. However, these technologies have an important drawback: the scope. RFID technology using passive tags, i.e. tags that don't need any external power supply, only can be read from a few centimeters. Although, recently a RFID reader has been developed by the Daily RFID company, which can be read passive tags from 15 meters. In this case, the main problem of the reader is its size, upper than 40 cm.

However, this technology has a crucial advantage over other identification methods: the cost. We will focus on the embedded RFID and NFC readers due to in the future, studying the market trends; most generation mobile phones could include this technology. This is possible thanks to the increase in electronic payment (Ondrus, 2007) (Lacmanovic, Radulovic, & Lacmanovic, 2010) and, as mentioned above, the reduced cost of both, readers and tags.

Internet of Things philosophy is based on that every object must have an electronic resource which al-lows identifying it. This resource must be composed by two elements: an identification method (in our case RFID or NFC technology) and an information provider. The last one is often a server with internet connection through which are offered all services to clients.

In this work we will develop a new way to present on the mobile phone services that smart-things give to users and interact with their environment. In this case, all the information needed by these services will be stored in the own mobile phone, being the charge of managing the user information and becoming in his/her Digital Me (Andreas Björklind, 2003). So, Digital Object Memory project (Boris Brandherm, 2010), in which objects have information about the user and his/her interaction with them, might improve its QoS (especially in terms of personalization) based on data from the mobile phone.

One step further, the proposed work will receive a feedback from Object Memories, with the aim of improve the QoS of local services.

Furthermore, this proposal not only allows sharing information between user's device and smart-thing, but the user can share all information obtained from the items with her/his friends, due to the system allows connections with social networks for this purpose. Additionally, the user can share some smart services with friends, and getting others to use services offered by these smart-things.

Due to the system make use of personal information, this must take in mind the aim of assure the information privacy. This will be made by use a core-application around which will be executed services and functions from smart-things. Thus, the user can be sure that only the needed information will be shared with the smart-thing and it's more, the own user will be able to determine what information wants to share and what not.

All these features will make the user can use the functionality provided for his/her environment in a more natural way and without interact directly with the smart-thing. Therefore the user must not learn how the item can be used, but only will have to watch the phone screen and execute the service.

1.2   Contributions of This Work

The novelty of this work relies on users must be execute and share services offered by smart-things that surround them, and this execution must be based on a minimum user interaction. To this end, we introduce in this paper a set of facilities and techniques that will make possible achieve the cited aim:

— Control of shared information. Mobile devices store much personal information about the user. Possibilities of share this information are huge, but must be present that an improper access to it could be fatal. For this reason, in this paper will be exposed a method to notify the user what in-formation is requested by each service. Further-more, security clearances will be stored in the own device, so the next time the user interacts with this service, the information will be shared automatically.

— Sharing services functionality. Users don't only can execute a service provided by some smart-thing, but also they will be able to share this service with any trusted people. This will be made by using social networks as sharing platform and delivering to their members the service address. As mentioned above, each service will have a concrete http address, so a call to this service by using the correct parameters will execute the functionality. For this purpose, has been developed a method to determine if some service is or not shareable and if a person has privileges to execute it.

— On-line and Off-line services. The proposed sys-tem has been made for allowing users select if must be executed the on-line or off-line version of some service. As we will see later, some service could be executed using a user-interface application or an automatic execution by web-services calls. The first one can be seen as a traditional application in which users must interact to for obtain a result. On the other hand, using on-line methods, the system delivers to users an autonomous process to send the information requested by some service and execute it to obtain results.

— Identification in On-line and Off-line services. Usually, the information sharing by on-line and off-line services will be different. In this way, while the on-line services can access to the user information stored, the off-line services must collect this information independently. So it's obvious that the functionality of both could be different. In this work is developed a method to show users what are differences between on-line and off-line versions of the same service. Once showed this information, the user can select what version wants to use.

— Improve QoS of local services with external data. In previous works, has been exposed some method to store information inside smart-things by using tags, Bluetooth and others technologies. However, is very interesting for the user using this information to improve the capabilities of the device. In this work, will be presented a new technique based on IoT and smart-things to improve the capabilities of the physical activity recognition of users (L. M. Soria-Morillo, 2010). These systems allow the device to know many physical activities carried out by the user, but the main problem of them is the limited number of activities recognized. The most of these systems use accelerometers to detect pattern associated to some physical activity.

— Improve QoS of remote process with local data. By storing the information on the own device, it's possible to share information between device and smart-thing. This information should be interesting for objects to improve de capabilities of them and theirs functionalities.

In the work presented along this paper, the main aim is to transforms the item-oriented view that Inter-net of Things has of the environment, towards a new concept in which the functionality is the cornerstone of the system. In this way, will be used a RESTful architecture to communicate the device (client-service) and the smart-thing access point (host-service). Later will be exposed the reasons why has been used this RESTful architecture for this purpose.

Furthermore, this change of view will facilitate the maintenance, scalability and developing of new services based on smart-things.

On the other hand, to share not only information but services with other users will let to get information and operate over remote objects thanks the access to its host-services elements.

1.3   Paper Structure

Our paper is structured as follows. In Section 2, a scenario will be exposed to show the functionalities of the system, in which a real situation problem will be solved by using the techniques developed in this paper.

Section 3 presents the main ideas behind the Inter-net of Things concept and technologies used to develop our system.

Later, in Section 4, will be exposed an overview of the proposal, with different modules that compose the system. Along this section will be introduced the ways to share information between these modules as well as the structure that will have the information stored both the smart-thing server and user's mobile device.

In Section 5 is exposed in a more detailed way the different methods to execute the services. In this section will both be compared to determine limitations of each ones. Finally, will be presented a method allowing to users make a feedback about the service executed and based on users community feedbacks will be developed an recommendation system. This advice sys-tem allows to the user obtain automatically the best service for each item based on his/her information stored and his/her user profile.

Section 6 contains the different ways through which users can interact with smart-things. Initially, the differences between these methods will be where the user is and the kind of interaction between the user's device and the tagged object.

Finally, Section 7 contains conclusions which can be reached after the completion of this work.

2   Scenario

Mary really loves new technologies and therefore often she buys gadgets that provide new features to its environment and help her to carry out the daily activities.

However, not everyone is happy at Mary's home. Daniel, her husband, seemed overwhelmed with the large number of devices in their home, so he generally made use of a few services, and is far from fully exploit all functionalities. But all that changed a few months ago with the arrival of the new detection and information sharing system.

The system, which provides the majority of devices at home of a recognition system based on NFC tags, make Mary and Daniel's mobile devices have become the main attraction of the house. In this way the system will be present at all times, inside and out-side the home.

Its 7:00 AM, the clock rings and Daniel is about to go at work. After picking up the phone from the table, our friend goes to the kitchen to drink a coffee cup made by him. To do this, Daniel bring his device close to the coffee maker, and immediately the device screen shows all functions offered by the machine (cleaning, make coffee, turn on, turn off, schedule, etc.). All these functions are obtained through the service interface provider by the manufacturer and this information is formatted by the mobile application. Our friend selects "make me my favorite coffee". Automatically, the service request information about Daniel's coffee tastes, which have been previously stored in the device. With this gesture, the coffee maker is able to make our friend's favorite coffee.

After taking the coffee, and automatically published this information through tweeter, Daniel is pre-pared to dress. To this end, the closet has a tag that, after being scanned, offers to Daniel different features, among which is "choosing clothes." Since Daniel device contains information about daily activities, the device sends to the service an indication that today is a business day and the time that Daniel chooses a formal costume. The suggested clothe based on weather expected for today is shown in Daniel's device.

After dressing and taking coffee, Daniel is ready to go at work, so down the stairs to the garage and ride in his car. At this time, due to localized services functionality, the option to open the garage door is displayed on the device. Once in the car, the device displays the option "search traffic incidences" to his work, allowing him to choose between an on-line application based on Google maps or activate the browser of his vehicle. Since the location of the work is stored on the device, the system automatically obtains all issues related to the route taken usually. Once checked that there is no traffic problems, starts his journey. Meanwhile in Facebook, the system automatically published that Daniel goes to work, and the status changed to "driving" warning to his friends that they should not call him.

The morning runs quietly while Mary, at home, prepares to listen to music. To do this action, Mary brings the device closer to the label on her Hi-Fi and chooses "default playlist". This option will allow to the Hi-Fi getting the most listened music in Mary's device and play at home stereo. However, Mary notes that the "play track" is a shared service provided by the system, and she publishes this service link among her social network. From this point, all her friends can control the music played in the Mary's Hi-Fi. Meanwhile, Daniel receives an alert on his device which indicates that Mary has shared a service with him. So without missing a beat, Daniel plays remotely the favorite song of both, a huge surprise for Mary.

3   Internet of Things Requirements

Here should be explained on what is basis on the Internet of Things and what we need to know about the item (smart-thing) to get a working system. Also, will be described in what consist on the RESTful architecture and what is the advantage of sharing the information managed in this work through this technique, instead of using other systems such as SOAP. It will also include REST verbs, i.e. functions GET, POST, PUT and DELETE, and how can we modeling services by using these functions.

In this way, application installed on smartphones must link smart items with on-line services associated. So should be noted that if some change in original application were required, manufacturer just need to change the service implementation and the associated XML services description.

NFC device is responsible for storage the basic XML descriptive file, so interaction between user's smartphone and smart item is quick and straight. In fact, this not only allow wired services connected to manufacturer or OMS server, but execute preloaded servicer or applications directly with no internet connection needed. In the evaluation process by users this feature was very interesting for them, particularly when smart-items were in non-urban areas such as fields, farmhouses or cottage without internet coverage.

4   Overall Process

Overview of the execution process of an application based on smart-thing as provider and how the device exchanging necessary information with the application access point. The latter is very interesting from the point of view that the user mobile phone is the information manager, so all requested information from smart-thing services must be provided by the user's device. If some information is not stored, application controller will request it to the user.

Flowchart describing this section is shown in *Figure 1*. In this figure can be seen different subsystems identified in the main application, which will be responsible for manage applications and services pro-vides by smart-items.

First, identification subsystem is presented. Smartphone reads NFC tag and obtains all information contained on it, which describes services provided by smart items.

Then, action set is shown to user, such us smart-item interaction history, information stored into (useful for active items) or services provided. In the latter option, two alternatives can be offered: on-line or off-line services. If the service requires information about user, the user must approve information exchange. This feature is implemented as a security mechanism for deny non-approved information exchange, which is critical when contextual or personal information is consumed.

If required information is not storage or it can't provide by any context provider, system will require it to user. Depending of information type, a form will be displayed and user could fill the information called for.

Once all information is stored, service or local application could be executed and result will be shown to the user.

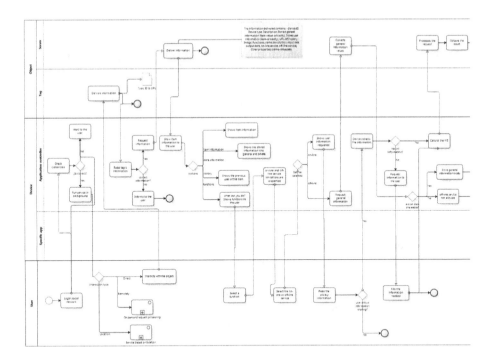

Fig. 1. Interaction flow diagram between users and Smart-Things

5   Services

Applications can provide some services to users for facilitate the use of smart-items or share information with them. In first place, must be distinguished two services kinds, on-line services implemented in the device and off-line services, implemented on the server. At this point we should explain what information is shown to the user to choose between on-line and off-line and, of course, what is the best method for interact with smart-items based on the environment.

For on-line services description, is needed to store a certain information kind to connect the application with a web service which provide the functionality. For this purpose has been defined a XML structure for describe de functionality and parameters required by this service.

The schema is composed by four big sections: item information block, services block, user interface configuration block and security assurance block. Now it will be described the content of previous blocks.

— Information block. Contains every kind of smart-thing static information, which could be understood as item specifications. This information is standardized into a specific ontology for allow queries and automatic information processing. Some information as location is provided in a dynamic way when objects are scanned by using NFC sensors embedded into smartphones. This means that must be needed to use two information block versions. First, a read-only version provided by OMS server, which contains common information for same type items. Second, a modified version of the previous document filled with concrete information of the labeled product, for instance, with product location or product owner.

— Services block. In this section is described the services offered by the product. Obviously, products can offers one or more services. Each service must contain general information about it, such as name, identification or description. At this point, must be noted that two kinds of services exist. On one hand, virtual service which results will be logic information, for instance, electronic program guide for some channel or temperature on the current city. On the other hand, real service could be implemented over multimedia objects. These services can change the real state for the object, such as change current program in a TV or change tuner on the radio. Regardless of the type of service, must be returned some value. A type set of values is defines in the structure for this purpose. In addition, often services need parameters for the execution. These parameters are defined in the structure and can be filled by the own user (basic input method) or obtained automatically from context providers (Luis M. Soria-Morillo, Juan A Ortega, 2011).

— User interface block. The aim of this section is to configure user interface appearance of different applications or services. Appearance scopes are not deeps and it just allows customizing concrete aspects. This decision was taken for achieve standard interfaces to make easier human-application interactions.

— Security assurance block. Due to it's possible to define location based services, which are activated when users are in a particular place. To avoid intrusions on personal systems, users can define services that just can be executed by a subset of users. For instance, it could be useful for allow TV tuning. If some neighbor is in our house, may not be interesting he/she can change the TV channel, but yes for our child. Furthermore, NFC technology provides the possibility of implementing a hardware security system by restricting the number of de-vices that can read a particular tag (Haselsteiner & Breitfuß, 2006).

5.1   Off-Line Services

These services are implemented like typical applications. These applications will be provided by the own smart-thing and could be installed in a transparent way in the user mobile phone just reading the associated tag. These applications can't access to the information stored on the device, due to could be break the security of this information.

5.2   On-Line Services

These services give to the user a new way to inter-act with smart-thing. Smart-things in this case, not only will provide information to the user, but will allow carrying out some processing. These processes are so called "services", and it's the main difference between passive objects and active objects. First ones only allow request information about the state, while the active objects allow to the user interact with them and request that some service (process) was made.

6   Interaction Methods

An interaction method is considered as a way in which a user can interact with his/her environment services. At first there are three ways of interaction, direct interaction and remote interaction and interaction by location.

— A direct interaction is presented when users maintain physical contact with items. This is the most trivial interaction and it's used most of the time nowadays.
— If the previous physical interaction is replaced by any smartphone with some application to control the smart-item we are facing to a re-mote interaction. This interaction kind allows delivering information needed by the function by querying the user preferences.
— Interaction by location is a new kind of inter-action system that allows publishing functionalities about smart-items just on a concrete location. It's to say, if some user is in the kitchen and a certain smart-item only offers some functionality in the living room, it won't be available where the user is.

7   Conclusions

This paper establishes the basis of a new system which will be able to share information in an automatic way between smart-items and user's smartphone. Furthermore, thanks to the proposal ontology, every item could be described by a set of properties which will be used for apply filters over the items and by this way, assigns a set of authorizations for share just some information about the user, if it is required by the application.

The proposed system will allow interacting with smart-items across a new concept: digital me. In this structure, whole information about user is stored in his/her smartphone, being that needed inputs for functions offered by items. Considering this concept, user does not have to introduce the needed information, but that will be obtained in an automatic way.

Obviously this is not a trivial approach, due to the quantity and kinds of scenarios are wild. For this purpose and to lead the way of the new system, a concrete scenario has been raised.

Acknowledgements. This research is supported by the Spanish Ministry of Science and Innovation R&D project ARTEMISA (TIN2009-14378-C02-01).

References

1. Alexander Kröner, A.J.: Augmenting Cognition With a Digital Episodic Memory, Künstliche Intelligenz, KI, pp. 51–57 (2008)
2. Andreas Björklind, S.H.: Ambient Intelligence to Go. Research Report SAR-03-03 (2003)
3. Boris Brandherm, J.H.: Demo: Authorized access on and interaction with Digital Product Memories. In: Eight Annual IEEE International Conference on Pervasive Computing and Communications, PerCom 2010, pp. 838–840. IEEE, Mannheim (2010)
4. von Reischach, F., Michahelles, F., Guinard, D., Adelmann, R., Fleisch, E., Schmidt, A.: An Evaluation of Product Identification Techniques for Mobile Phones. In: Gross, T., Gulliksen, J., Kotzé, P., Oestreicher, L., Palanque, P., Prates, R.O., Winckler, M. (eds.) INTERACT 2009. LNCS, vol. 5726, pp. 804–816. Springer, Heidelberg (2009)
5. Kranz, M., Holleis, P., Schmidt, A.: Embedded Interaction: Interacting with the Internet of Things. IEEE Internet Computing, 46–53 (2010)
6. Soria-Morillo, L.M.,, J.A.-G.-A.: Tracking system based on accelerometry for users with restricted physical activity. In: 23rd International Conference on Industrial Engineering and Others Applications of Applied Intelligent Systems, iea/aie 2010, Córdoba, Spain (2010)
7. Lacmanovic, I., Radulovic, B., Lacmanovic, D.: Contactless payment systems based on RFID technology. In: 2010 Proceedings of the 33rd International Convention MIPRO, pp. 1114–1119. IEEE Computer Society, Opatija (2010)
8. Ondrus, J.P.: An Assessment of NFC for Future Mobile Payment Systems. In: Conference on the Management of Mobile Business, p. 43. IEEE Computer Society, Washington, DC (2007)
9. Roy, W.,, F.K.: Bridging physical and virtual worlds with electronic tags. In: CHI 1999: Proceedings of the SIGCHI Conference on Human Factors in Computing Systems, pp. 370–377. ACM, New York (1999)

A Conversation Model Enabling Intelligent Agents to Give Emotional Support

Janneke M. van der Zwaan[1], Virginia Dignum[1], and Catholijn M. Jonker[2]

[1] ICT Section, Delft University of Technology, Jaffalaan 5, Delft, The Netherlands
j.m.vanderzwaan@tudelft.nl
[2] Interactive Intelligence group, Delft University of Technology, The Netherlands

Abstract. In everyday life, people frequently talk to others to help them deal with negative emotions. To some extent, everybody is capable of comforting other people, but so far conversational agents are unable to deal with this type of situation. To provide intelligent agents with the capability to give emotional support, we propose a domain-independent conversational model that is based on topics suggested by cognitive appraisal theories of emotion and the 5-phase model that is used to structure online counseling conversations. The model is implemented in an embodied conversational agent called Robin.

1 Introduction

To alleviate stress and deal with negative emotions, people frequently turn to others to talk about their problems. Most people are more or less successful in comforting others. Early work in the field of affective computing demonstrated that virtual agents are able to reduce negative emotions in users [5]. More recent developments show that empathic agents are increasingly capable of complex social and emotional dialogues, but so far they are unable to comfort users.

We are interested in investigating how and to what extent conversational agents can provide social support. Social support or comforting refers to communicative attempts to alleviate the emotional distress of another person [3]. Our research concerns the design and evaluation of an Embodied Conversational Agent (ECA) that provides social support to victims of cyberbullying [11]. Cyberbullying is bullying through electronic communication devices [6]. Although in our application we assume the user's problem is connected to cyberbullying, we want to keep the strategies for giving social support as domain-independent as possible.

General advice on comforting includes letting the other party talk about their 'thoughts and feelings' regarding the upsetting situation [10]. However, this does not specify what topics might be addressed. To overcome this gap, Burleson and Goldsmith linked comforting to cognitive appraisal theories of emotions and identified a sequence of topics in comforting conversations [3]. Our agent uses this model to determine what topic it will bring up next. The goal of this paper is to present a domain-independent conversational model of comforting. We also introduce our implementation of this model in a conversational agent called Robin.

The agent and the user communicate predominantly through natural language text messages. Given the complexity of interpreting and generating natural language, in the

W. Ding et al. (Eds.): Modern Advances in Intelligent Systems and Tools, SCI 431, pp. 47–52.
springerlink.com © Springer-Verlag Berlin Heidelberg 2012

current system, the text interpretation and generation have not been implemented. Instead, we use logical representations of utterances (speech acts). This abstraction allows us to focus on the agent's reasoning process and show how it can use appraisal theory to give social support.

This paper is organized as follows. In the next section, we provide a background on emotion theories. In Sect. 3, we present the conversation model. Section 4 introduces the comforting agent that implements the conversation model. Section 5 presents our conclusions.

2   Background

To assist users in dealing with their negative emotions, the agent needs be aware of the user's emotional state and the reasons these emotions arise. Within the field of psychology different emotional theories exist. Two major strands of emotion theories that can be distinguished are dimensional theories [4] and cognitive appraisal theories of emotion [8].

Dimensional emotion theories are based on the idea of classifying emotions along an arbitrary amount of dimensions of connotative meaning. A well known dimensional model is the Pleasure, Arousal, Dominance (PAD) model of emotions, which assumes an emotion (more precisely: affect) can be defined as a coincidence of values on the dimensions pleasure, arousal and dominance [7,9].

The basic assumption of cognitive appraisal theories is that emotions are triggered by the evaluation of a stimulus with respect to several criteria, including personality dimensions, characteristics of the situation, and other context variables [8]. The evaluation of a stimulus is called appraisal. Different cognitive theories of appraisal exist, they differ mostly in the dimensions that make up the appraisal process. Ortony, Clore and Collins proposed a computationally friendly formalization of cognitive appraisal theories [8]. The so-called OCC model is frequently used as a model for computerized emotion.

Burleson and Goldsmith use Lazarus' cognitive appraisal theory of emotions to explain how the process of comforting works [3]. After observing emotions are elicited by appraisals of situations and not by the situations themselves, they conclude that a distressed emotional state can only be altered by changing the appraisals that underlie the negative emotion [3]. Helpers can facilitate these reappraisals by encouraging the distressed person to explore and clarify thoughts and feelings that are relevant to the stressful situation. As a result, the distressed person may change his goals, views of the situation, and/or coping efforts [3].

Based on these observations, Burleson and Goldsmith suggest a set of topics that can be addressed in the course of a supportive conversation. First, the conversation should focus on getting a fuller appreciation of the emotional state, the reasons for its occurrence, and an assessment of its appropriateness. After that, the current coping strategy and its effectiveness can be discussed. Finally, if the current coping strategy does not seem to solve the problem, alternative coping strategies can be explored [3].

To further structure the conversation between the agent and the user, the conversational topics identified by Burleson and Goldsmith will be linked to phases in the

5-phase model. The 5-phase model was developed as a methodology to structure counseling conversations via telephone and chat [1]. The five phases of a conversation are: 1) Warm welcome: the counselor connects with the child; 2) Gather information: the counselor tries to establish the problem of the child; 3) Determine the objective of the session: the counselor and the child determine the goal of the conversation (e.g., getting tips on how to deal with bullying); 4) Work out the objective: the counselor stimulates the child to come up with a solution; and 5) Round off: the counselor ends the conversation.

3   Conversation Model

To get a model for comforting conversations, we combined the structure of the 5-phase model with the topics Burleson and Goldsmith suggested based on appraisal theory. Table 1 shows the division of the topics among the phases. In the explanation below, the term comforter refers to the agent and the distressed person is the user.

Table 1. Conversation phases and topics

Conversation phase	Topic(s)
1) Welcome	Hello
2) Gather information	Event (general)
	Emotional state
	Personal goal
	Event (details)
	Coping (current) (if need to cope)
3) Determine conversation objective	Conversation objective
4) Work out objective	Coping (future) (if need to cope)
	Advice (depending on conversation objective)
5) Round off	Bye

In the first conversation phase (welcome) the comforter greets the distressed person. In phase 2 (gather information), the comforter first asks the distressed person questions to identify facts about the event that elicited emotions. The second topic is the distressed person's emotional state. To be able to determine whether the distressed person needs to cope, the comforter also registers the intensity of the emotion. If the emotional state is negative, appraisal theory suggests a personal goal is threatened. Subsequently, the comforter asks the distressed person to specify which goal is threatened. After that, the comforter asks more details about the event. These details (together with information already acquired) determine the advice the comforter can give in phase 4. Which details are relevant and how details are mapped to advice depends on the domain in which the agent operates. When the comforter has gathered sufficient information (what is sufficient also depends on domain knowledge) and believes the distressed person needs to cope, it moves on to discussing the distressed person's current coping strategy. Then phase 3 (determine conversation objective) starts and the comforter asks the distressed person what he or she wants to accomplish with the conversation. For simplicity we assume the distressed person either wants to tell his story or get advice to deal with the problem. For both conversation objectives, phase 4 (work out objective) starts with discussing the distressed person's future coping options. If the distressed person wants advice on how to deal with his problem, subsequently the comforter gives advice. Finally, in phase 5 (round off), the comforters says goodbye to the distressed person.

4   Robin, the Comforting Agent

The conversation model introduced in the previous section was implemented in an embodied conversational agent called Robin. Robin tries to empower victims of cyberbullying by giving emotional support and practical advice. Figure 1 shows a screenshot of the system. The user communicates with Robin through a chat interface. As mentioned before, in the current implementation, verbal communication consists of speech acts instead of natural language utterances. The agent's embodiment consists of a set of pictures of the iCat (top left of Fig. 1). To communicate his emotional state, the user manipulates the AffectButton [2] (bottom left of Fig. 1). The AffectButton is a button with a rudimentary and gender-neutral face that changes its expression based on the position of the mouse cursor. By clicking the button when it shows the emotional expression the user wants to communicate, the corresponding PAD values are send to the agent. The emotion input is further explained in Sect. 4.1.

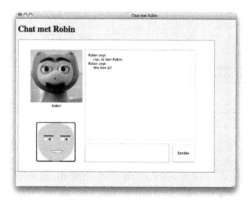

Fig. 1. Screenshot of Robin, the comforting agent

The input of the agent consists of speech acts and PAD values. The output consists of speech acts and one of five discrete emotional expressions (happy, surprised, sad, angry, afraid).

4.1   Emotion Input

To gather information about the user's current emotional state, Robin uses the Affect-Button (see Fig. 1). The AffectButton is a tool for explicit affective feedback. Since the user's emotional state is explicitly being addressed during the conversation, explicit emotion input arguably is more suitable than implicit emotion input (e.g., by recognizing facial expressions), because it requires the user to consciously think about his emotions. The AffectButton sends PAD triplets to the agent's reasoning engine.

To be able to interpret emotional states, PAD triplets have to be mapped to OCC emotion types. Since PAD and OCC are based on two different strands of emotion theories, no fixed mapping between these two models exists. Therefore, we propose a generic mapping. Based on the assumption that the user is dealing with a negative *event* all input emotions are mapped to well-being emotions. The value for pleasure determines whether an emotion is positive or negative; positive emotions ($P \geq 0$) are mapped to the emotion type joy and negative emotions ($P < 0$) are mapped to distress. The value of P also determines the intensity of the emotion: $-1 \leq P < -0.66$ or $0.66 < P \leq 1$, high intensity; $-0.66 \leq P < -0.33$ or $0.33 < P \leq 0.66$, medium intensity; $-0.33 \leq P < 0$ or $0 \leq P \leq 0.33$, low intensity. The agent assumes coping is necessary for medium and high negative emotions.

4.2 Emotion Output

Emotion output is realized by the embodiment and consists of one of five discrete emotional expressions (happy, surprised, sad, angry, afraid). The default expression for the agent is happy. Each time the user communicates an emotional expression with the AffectButton, the agent verbally confirms and mirrors this emotional state. After confirming the user's emotional state, the facial expression changes back to its default expression (happy).

5 Conclusion

In this paper, we presented a domain-independent dialogue model for comforting conversations. The model combines the 5-phase model, used to structure counseling conversations, with topics that are relevant according to cognitive appraisal theories of emotion. We also introduced our implementation of this model in a conversational agent that provides social support to victims of cyberbullying. The agent uses the AffectButton to obtain information about the user's emotional state. The AffectButton is based on a dimensional emotion model, whereas the conversational model is based on appraisal theories. Another contribution of this paper is a mapping from PAD values to OCC emotion types. The next step is to evaluate the agent. We want to determine to what extent conversations generated by the model are being perceived as supportive by users. Since we are dealing with a sensitive topic (cyberbullying) and a vulnerable target audience (children and adolescents), initially the agent will be evaluated by experts.

References

1. de Beyn, A.: In gesprek met kinderen:de methodiek van de kindertelefoon. In: SWP (2003)
2. Broekens, J., Brinkman, W.P.: Affectbutton: Towards a standard for dynamic affective user feedback. In: Affective Computing and Intelligent Interaction, ACII 2009 (2009)

3. Burleson, B.R., Goldsmith, D.J.: How the Comforting Process Works: Alleviating Emotional Distress through Conversationally Induced Reappraisals. In: Handbook of Communication and Emotion: Research, Theory, Applications, and Contexts, pp. 245–280. Academic Press, San Diego (1998)

4. Gehm, T.L., Scherer, K.R.: Factors determining the dimensions of subjective emotional space. In: Facets of Emotion: Recent Research, pp. 99–113. Lawrence Erlbaum Associates, Inc., Hillsdale (1988)

5. Hone, K.: Empathic agents to reduce user frustration: The effects of varying agent characteristics. Interact. Comput. 18(2), 227–245 (2006)

6. Li, Q.: New bottle but old wine: A research of cyberbullying in schools. Computers in Human Behavior 23(4), 1777–1791 (2007)

7. Mehrabian, A.: Basic Dimensions for a General Psychological Theory. Oelgeschlager, Cambridge (1980)

8. Ortony, A., Clore, G.L., Collins, A.: The cognitive structure of emotions. Cambridge University Press, Cambridge (1988)

9. Osgood, C.E.: Dimensionality of the semantic space for communication via facial expressions. Scandinavian Journal of Psychology 7(1), 1–30 (1966)

10. Pennebaker, J.W.: Putting stress into words: Health, linguistic, and therapeutic implications. Behaviour Research and Therapy 31(6), 539–548 (1993)

11. van der Zwaan, J.M., Dignum, V., Jonker, C.M.: Simulating peer support for victims of cyberbullying. In: Proceedings of the 22st Benelux Conference on Artificial Intelligence, BNAIC 2010 (2010)

Rationality for Temporal Discounting, Memory Traces and Hebbian Learning

Jan Treur[*] and Muhammad Umair

VU University Amsterdam, Agent Systems Research Group
De Boelelaan 1081, 1081 HV Amsterdam, The Netherlands
COMSATS Institute of Information Technology, Lahore, Pakistan
treur@cs.vu.nl, mumair@ciitlahore.edu.pk
http://www.few.vu.nl/~treur

Abstract. In this paper three adaptive agent models incorporating triggered emotional responses are explored and evaluated on their rationality. One of the models is based on temporal discounting second on memory traces and the third one on hebbian learning with mutual inhibition. The models are assessed using a measure reflecting the environment's behaviour and expressing the extent of rationality. Simulation results and the extents of rationality of the different models over time are presented and analysed.

Keywords: adaptive agent model, memory traces, temporal discounting, mutual inhibition, Hebbian learning, rationality.

1   Introduction

Adaptive agents develop decision making capabilities over time based on experiences with their environment. In order to do so, such agents exploit certain learning mechanisms, for example, involving emotional responses triggered by situations. By the learning processes the decision making is adapted to the characteristics of the environment, so that the decisions made are in some way rational, given the enviroment as reflected in the agent's experiences. In this paper the focus is on three of such learning mechanisms: temporal discounting, memory traces and Hebbian learning with inhibition. The adaptation model based on *temporal discounting* adapts connections based on the frequency of occurrence of certain situations, thereby valuing occurences further back in time lower. A second alternative considered is a case-based memory modelling approach based on *memory traces* e.g. [20, 22]. For the Hebbian learning different variations are considered, by applying it to different types of connections in the decision model.

In this paper, in Section 2 the basic agent model is introduced. Section 3 and Section 4 presents the adaptation model based on temporal discounting and memory traces respectively. In Section 5 the adaptation model based on Hebbian learning with mutual inhibition is introduced. Simulation results are presented in a separate

---

[*] Corresponding author.

W. Ding et al. (Eds.): Modern Advances in Intelligent Systems and Tools, SCI 431, pp. 53–61.

Appendix A. In Appendix B two measures for rationality used are presented, and the different adaptation models are evaluated based on these measures.[1] Finally, Section 6 is a discussion.

2   The Basic Agent Model Used

This section describes the basic agent model in which three adaptation models discussed in Section 3, Sections 4 and 5 are incorporated. It is an extension of the work presented in [21]; see Figure 1 for an overview. It is assumed that responses in relation to a sensory representation state roughly proceed according to the following causal chain for a *body loop* (based on elements from [4, 7, 8]). An *as-if body loop* uses a direct causal relation from preparation for bodily response to sensory representation of body state; cf. [7]. This can be considered a prediction of the action effect by internal simulation (e.g., [13]). The resulting induced feeling provides an emotion-related valuation of this prediction (cf. [1, 2, 14, 15, 17]). If the level of the feeling (which is assumed positive here) is high, a positive valuation is obtained. The body loop (or as-if body loop) is extended to a recursive (as-if) body loop by assuming that the preparation of the bodily response is also affected by the level of the induced feeling; cf. [8], pp. 91-92. In this way the emotion-related valuation of the prediction affects the preparation. To adequately formalise this the hybrid dynamic modelling language LEADSTO has been used; cf. [4,5]. Fig. 1 also shows representations from the detailed specifications explained below, through local properties LP0 to LP6.

Note that the effector state for b_i (bodily responses) combined with the (stochastic) effectiveness of executing b_i in the world (indicated by *effectiveness rate λ_i* between *0* and *1*) activates the sensor state for b_i via body loop as described above. By a recursive as-if body loop each of the preparations for b_i generates a level of feeling for b_i which is considered a valuation of the prediction of the action effect by internal simulation. This in turn affects the level of the related action preparation for b_i. Dynamic interaction within these loops results in equilibrium for the strength of the preparation and of the feeling, and depending on these values, the action is actually activated with a certain intensity. The specific strengths of the connections from the sensory representation to the preparations, and within the recursive as-if body loops can be innate, or are acquired during lifetime. In this paper the considered adaptation mechanisms for the model are based on Hebbian learning with mutual inhibition, temporal discounting, and memory traces (Section 3). The detailed specification of the basic model is presented below starting with how the world state is sensed.

LP0 Sensing a world state

If   world state property w occurs of level V then the sensor state for w will have level V.

world_state(W, V) $\rightarrow$ sensor_state(W, V)

---

[1] These appendices can be found at URL
  http://www.few.vu.nl/~wai/IEA12rationality

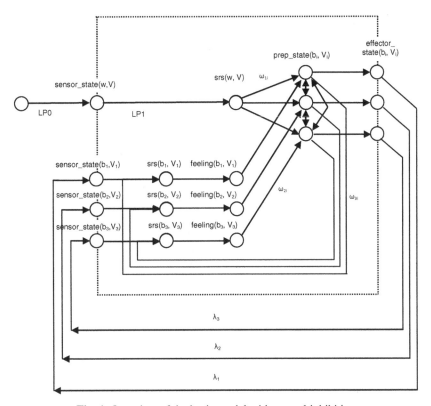

Fig. 1. Overview of the basic model with mutual inhibition

From the sensor state a sensory representation of the world state is generated by dynamic property LP1.

LP1 Generating a sensory representation for a sensed world state
If   the sensor state for world state property w has level V,
then   the sensory representation for w will have level V.
  sensor_state(W, V) → srs(W, V)

The function h combining two inputs activating a subsequent state is used along with a threshold function th to keep the value in the interval [0, 1] as follows: $h(\sigma, \tau, V_1, V_2, \omega_1, \omega_2) = th(\sigma, \tau, \omega_1 V_1 + \omega_2 V_2)$ where V_1 and V_2 are the activation level of the states and ω_1 and ω_2 the connection strengths of the links between the states; here

$$th(\sigma, \tau, V) = \left(\frac{1}{1+e^{-\sigma(V-\tau)}} - \frac{1}{1+e^{\sigma\tau}}\right)(1 + e^{-\sigma\tau})$$

where σ is the steepness and τ is the threshold of the given function. Dynamic property LP2 describes the generation of the preparation state from the sensory representation of the world state and the feeling.

LP2 From sensory representation and feeling to preparation of a body state
If   a sensory representation for w with level V occurs
   and the feeling associated with body state b_i has level V_i
   and the preparation state for b_i has level U_i
   and ω_{1i} is the strength of the connection from sensory representation for w to preparation for b_i
   and ω_{2i} is the strength of the connection from feeling of b_i to preparation for b_i
   and σ_i is the steepness value for preparation of b_i
   and τ_i is the threshold value for preparation of b_i
   and γ_1 is the person's flexibility for bodily responses
then   after Δt the preparation state for body state b_i will have
        level $U_i + \gamma_1\ (h(\sigma_i,\ \tau_i,\ V,\ V_i,\ \omega_{1i},\ \omega_{2i}\ )- U_i)\ \Delta t.$
   srs(w, V) & feeling(b$_i$, V$_i$) & preparation_state(b$_i$, U$_i$) &
   has_connection_strength(srs(w), preparation(bi), ω_{1i}) &
   has_connection_strength(feeling(b$_i$), preparation(b$_i$), ω_{2i}) &
   has_steepness(prep_state(b$_i$), σ_i) & has_threshold(prep_state(b$_i$), τ_i)
   $\rightarrow$ preparation(b$_i$, U$_i$ + γ_1 (h(σ_i, τ_i, V, V$_i$, ω_{1i}, ω_{2i},)- U$_i$) Δt)

Dynamic property LP3 describes the generation of sensory representation of a body state from the respective preparation state and sensory state.

LP3 From preparation and sensor state to sensory representation of a body state
If   preparation state for b_i has level X_i
   and sensor state for b_i has level V_i
   and the sensory representation for b_i has level U_i
   and ω_{3i} is the strength of the connection from preparation state for b_i to sensory representation for b_i
   and σ_i is the steepness value for sensory representation of b_i
   and τ_i is the threshold value for sensory representation of b_i
   and γ_2 is the person's flexibility for bodily responses
then   after Δt the sensory representation for body state b_i will have
        level $U_i + \gamma_2\ (h(\sigma_i,\ \tau_i,\ X_i,\ V_i,\ \omega_{3i},\ 1)- U_i)\ \Delta t.$
   preparation_state (b$_i$, X$_i$) & sensor_state(b$_i$, V$_i$) & srs (b$_i$, U$_i$) &
   has_connection_strength(preparation(bi), srs(bi), , ω_{3i}) & has_steepness(srs(b$_i$), σ_i) &
   has_threshold(srs(b$_i$), τ_i)
   $\rightarrow$ srs(b$_i$, U$_i$ + γ_2 (h(σ_i, τ_i, X$_i$, V$_i$, ω_{3i}, 1)- U$_i$) Δt)

Dynamic property LP4 describes how the feeling is generated from the sensory representation of the body state.

LP4 From sensory representation of a body state to feeling
If   the sensory representation for body state b_i has level V,
then b_i will be felt with level V.
   srs(bi, V) $\rightarrow$ feeling(bi, V)

LP5 describes how an effector state is generated from respective preparation state.

LP5 From preparation to effector state

If   the preparation state for b_i has level V,

then   the effector state for body state b_i will have level V.

  preparation_state(bi, V) $\rightarrow$ effector_state(bi, V)

LP6 describes how a sensor state is generated from an effector state.

LP6 From effector state to sensor state of a body state

If   the effector state for b_i has level V_i,

  and λ_i is world characteristics/ recommendation for the option b_i

then   the sensor state for body state b_i will have level $\lambda_i V_i$

  effecor_state(bi,V) & has_contribution(effecor_state(bi), sensor_state(bi), λ_i)

  $\rightarrow$ effector_state(bi, V) $\rightarrow$ sensor_state(bi, $\lambda_i V_i$)

For the case studies addressed three options are assumed available in the world for the agent and the objective is to see how rationally an agent makes decisions using a given adaptation model (under static as well as stochastic world characteristics).

3   The Three Adaptation Models

In this section the model is extended with the ability of the adaptation of the links (ω_{1i}, ω_{2i}, and ω_{3i}; see Fig. 1) among different states. The dynamic properties LP7 to LP9 describe the mechanism of learning of the connection strengths between sensory representation of the world state to preparation, feeling to preparation, and preparation to sensory representation, respectively. Appendix B shows some simulation results in detail. First the adaptation model based on the temporal discounting approach Is described.

LP7a Temporal discounting learning rule for sensory representation of world state

If   the connection from sensory representation of w to preparation of b_i has strength ω_{1i}

    and   the sensory representation for w has level V and $V>0$

    and   the discounting rate from sensory representation of w to preparation of b_i is η_i

    and   the extinction rate from sensory representation of w to preparation of b_i is ζ_i

then   after Δt the connection from sensory representation of w to preparation of b_i will have strength $\omega_{1i} +$
    ($\eta_i(V- \omega_{1i})- \zeta_i \omega_{1i}) \Delta t$.

  has_connection_strength(srs(w), preparation(b₁), ω_{1i}) & srs(w, V) & V>0 &

  has_discounting_rate(srs(w), preparation(b₁), η_i) &

  has_extinction_rate(srs(w), preparation(b), ζ_i)

  $\rightarrow$ has_connection_strength(srs(w), preparation(b₁), $\omega_{1i} + (\eta_1 (V- \omega_{1i}) - \zeta_i \omega_{1i}) \Delta t$)

LP7b Temporal discounting learning rule for sensory representation of world state

If   the connection from sensory representation of w to preparation of b_i has strength ω_{1i}

  and   the sensory representation for w has level 0

  and   the extinction rate from sensory representation of w to preparation of b_i is ζ_i

then   after Δt the connection from sensory representation of w to preparation of b_i
        will have strength $\omega_{1i} - \zeta_i \, \omega_{1i} \, \Delta t$.
    has_connection_strength(srs(w), preparation(b$_i$), ω_1) & srs(w, 0) &
    has_extinction_rate(srs(w), preparation(b$_i$), ζ_i)
    → has_connection_strength(srs(w), preparation(b$_i$), $\omega_{1i} - \zeta_i \, \omega_{1i} \, \Delta t$)

Similarly LP8 and LP9 specify the same temporal discounting mechanism for the connection from feeling to preparation, and from preparation to sensory representation.

Next the adaptation model based on memory traces is described. Here the dynamic properties LP7 to LP9 describe the mechanism of learning of the connection strengths using this a memory traces approach.

LP7a Discounting memory traces from srs(w) to prep(bi)

If   the sensory representation for w has strength V
 and   the preparation of bi has strength V_i
 and   the discounted number of memory traces with state srs(w) are X
 and   the discounted number of memory traces with state srs(w) and successor state preparation(bi) is Y
 and   the discounting rate from sensory representation of w to preparation of "bi" is α_i
 and   the extinction rate from srs(w) to preparation of bi is ζ_i
then   the discounted number of memory traces with state srs(w) is $X + \alpha_i V - \zeta_i X$
 and   the discounted number of memory traces with state srs(w) and successor state preparation(bi)
        is $Y + \alpha_i V V_i - \zeta_i Y$
    srs(w, V) & preparation(b$_i$, V$_i$) & has_discounting_rate(srs(w), preparation(b$_i$), α_i) &
    has_extinction_rate(srs(w), preparation(b$_i$), ζ_i) & memory_traces_including(srs(w), X) &
    memory_traces_including_both(srs(w), preparation(b$_i$), Y)
    → memory_traces_including(srs(w), X + α_i V - ζ_i X) &
        memory_traces_including_both(srs(w), preparation(b$_i$), Y + α_i VV$_i$ - ζ_i Y)

Given these numbers the induction strength of the connection from sensory representation to preparation state is determined as Y/X.

LP7b Generation of preparations(bi) based on discounted memory traces

If   the discounted number of memory traces with state srs(w) is X
 and   the discounted number of memory traces with state srs(w) and successor state preparation(bi) is Y
then   the connection strength from srs(w) to preparation(bi) is Y/X
    memory_traces_including(srs(w), X) &
    memory_traces_including_both(srs(w), preparation(bi), Y)
    → has_connection_strength(srs(w), preparation(bi), Y/X)

Similarly LP8 and LP9 specify the same memery trace mechanism for the connection from feeling to preparation, and from preparation to sensory representation.

Finally the adaptation model based on Hebbian learning with inhibition is described. Here the basic agent model described in Section 2 is extended with two additional features. One is mutual inhibition for preparation states and the other one is adaptation of connection strength using a Hebbian learning approach. Later in this section the overview of the model is discussed with detailed specifications in LEADSTO. An overview of the extended model for the generation of emotional responses and feelings is depicted in Figure. 1 with the extension of *mutual inhibition* at preparation states. In the current section only the local property LP2 is discussed which incorporate the effect of mutual inhibition in calculation of the value of the preparation state. Moreover, local properties LP7 to LP9 define the Hebbian learning approach. Dynamic property LP2 describes the generation of the preparation state from the sensory representation of the world state and the feeling thereby taking into account mutual inhibition. For this particular case the combination function is defined as:

$$g(\sigma, \tau, V_1, V_2, V_3, V_4, \omega_1, \omega_2, \theta_1, \theta_2) = th(\sigma, \tau, \omega_1 V_1 + \omega_2 V_2 + \theta_1 V_3 + \theta_2 V_4)$$

where θ_{mi} is the strengths of the mutual inhibition link from preparation state for b_m to preparation state for b_i (which have negative values).

LP2 From sensory representation and feeling to preparation with mutual inhibition

If   a sensory representation for w with level V occurs

and   the feeling associated with body state b_i has level V_i

and   the preparation state for each b_m has level U_m

and ω_{1i} is the strength of the connection from sensory representation for w to preparation for b_i

and ω_{2i} is the strength of the connection from feeling of b_i to preparation for b_i

and θ_{mi} is for each m the strength of the inhibition connection from preparation state for b_j

          to preparation for b_i

and σ_i is the steepness value for preparation of b_i

and τ_i is the threshold value for preparation of b_i

and γ_1 is the person's flexibility for bodily responses

and $j{\neq}i, k{\neq}i, j{\neq}k$

then   after Δt the preparation state for body state b_i will have

          level $U_i + \gamma_1 (g(\sigma_i, \tau_i, V, V_i, U_j, U_k, \omega_{1i}, \omega_{2i}, \theta_{ji}, \theta_{ki}) - U_i) \Delta t$

     srs(w, V) & feeling(b_i, V_i) & $\wedge_m$ preparation_state(b_m, U_m) &

     has_connection_strength(srs(w), preparation(bi), ω_{1i}) &

     has_connection_strength(feeling(b_i), preparation(b_i), ω_{2i}) &

     has_steepness(prep_state(b_i), σ_i) & has_threshold(prep_state(b_i), τ_i) & j≠i, k≠i, j≠k

     → preparation(b_i, $U_i + \gamma_1 (g(\sigma_i, \tau_i, V, V_i, U_j, U_k, \omega_{1i}, \omega_{2i}, \theta_{ji}, \theta_{ki}) - U_i) \Delta t$)

Moreover, the connection strength of the different links using Hebbian learning are updated according to the local properties LP7 to LP9.

LP7 Hebbian learning for connection from sensory representation of stimulus to preparation

If   the connection from sensory representation of w to preparation of b_i has strength ω_{1i}

   and   the sensory representation for w has level V

   and   the preparation of bi has level V_i

   and   the learning rate from sensory representation of w to preparation of b_i is η

   and   the extinction rate from sensory representation of w to preparation of b_i is ζ

then   after Δt the connection from sensory representation of w to preparation of b_i will have strength $\omega_{1i} +$

          $(\eta V V_i (1 - \omega_{1i}) - \zeta \omega_{1i}) \Delta t$.

has_connection_strength(srs(w), preparation(b_i), ω_{1i}) & srs(w, V) & preparation(b_i, V_i) &
has_learning_rate(srs(w), preparation(b_i), η) &
has_extinction_rate(srs(w), preparation(b_i), ζ)
$\rightarrow$ has_connection_strength(w, b_i, ω_{1i} + ($\eta VV_i (1 - \omega_{1i})$ - $\zeta\omega_{1i}$) Δt)

Similarly LP8 and LP9 specify the same Hebbian learning mechanism for the connection from feeling to preparation, and from preparation to sensory representation. In Appendix A an overview of simulation results is shown and Appendix B presents the evaluation of the models on rationality. These appendices can be found at http://www.few.vu.nl/IEA12rationality.

4   Discussion

This paper focused on three different adaptive agent models: the first based on temporal discounting, the second on memory traces and the third based on Hebbian learning with mutual inhibition; cf. [10, 12]. For all described adaptive models an analysis of the extent of rationality was made. The basic agent model in which adaptation models were incorporated is based on emotion-related valuation of predictions, involving feeling states generated in the amygdala; e.g., [1, 2, 6, 8, 14, 15, 17].

The assessment of the extent of rationality with respect to given world characteristics, was based on two measures. It was shown how by the learning processes indeed a high level of rationality was obtained, and how after major world changes with some delay this rationality level is re-obtained. It turned out that emotion-related valuing of predictions in the amygdala as a basis for adaptive decision making according to temporal discounting, to memory traces and to Hebbian learning all satisfy reasonable rationality measures.

References

1. Bechara, A., Damasio, H., Damasio, A.R.: Role of the Amygdala in Decision-Making. Ann. N.Y. Acad. Sci. 985, 356–369 (2003)
2. Bechara, A., Damasio, H., Damasio, A.R., Lee, G.P.: Different Contributions of the Human Amygdala and Ventromedial Prefrontal Cortex to Decision-Making. Journal of Neuroscience 19, 5473–5481 (1999)
3. Bosse, T., Hoogendoorn, M., Memon, Z.A., Treur, J., Umair, M.: An Adaptive Model for Dynamics of Desiring and Feeling Based on Hebbian Learning. In: Yao, Y., Sun, R., Poggio, T., Liu, J., Zhong, N., Huang, J. (eds.) BI 2010. LNCS, vol. 6334, pp. 14–28. Springer, Heidelberg (2010)
4. Bosse, T., Jonker, C.M., Treur, J.: Formalisation of Damasio's Theory of Emotion, Feeling and Core Consciousness. Consciousness and Cognition 17, 94–113 (2008)
5. Bosse, T., Jonker, C.M., van der Meij, L., Treur, J.: A Language and Environment for Analysis of Dynamics by Simulation. Intern. J. of AI Tools 16, 435–464 (2007)
6. Damasio, A.: Descartes' Error: Emotion, Reason and the Human Brain. Papermac, London (1994)

7. Damasio, A.: The Feeling of What Happens. Body and Emotion in the Making of Consciousness. Harcourt Brace, New York (1999)
8. Damasio, A.: Looking for Spinoza. Vintage books, London (2004)
9. Forgas, J.P., Laham, S.M., Vargas, P.T.: Mood effects on eyewitness memory: Affective influences on susceptibility to misinformation. Journal of Experimental Social Psychology 41, 574–588 (2005)
10. Gerstner, W., Kistler, W.M.: Mathematical formulations of Hebbian learning. Biol. Cybern. 87, 404–415 (2002)
11. Ghashghaei, H.T., Hilgetag, C.C., Barbas, H.: Sequence of information processing for emotions based on the anatomic dialogue between prefrontal cortex and amygdala. Neuroimage 34, 905–923 (2007)
12. Hebb, D.O.: The Organization of Behaviour. John Wiley & Sons, New York (1949)
13. Hesslow, G.: Conscious thought as simulation of behaviour and perception. Trends Cogn. Sci. 6, 242–247 (2002)
14. Montague, P.R., Berns, G.S.: Neural economics and the biological substrates of valuation. Neuron 36, 265–284 (2002)
15. Morrison, S.E., Salzman, C.D.: Re-valuing the amygdala. Current Opinion in Neurobiology 20, 221–230 (2010)
16. Murray, E.A.: The amygdala, reward and emotion. Trends Cogn. Sci. 11, 489–497 (2007)
17. Rangel, A., Camerer, C., Montague, P.R.: A framework for studying the neurobiology of value-based decision making. Nat. Rev. Neurosci. 9, 545–556 (2008)
18. Salzman, C.D., Fusi, S.: Emotion, Cognition, and Mental State Representation in Amygdala and Prefrontal Cortex. Annu. Rev. Neurosci. 33, 173–202 (2010)
19. Sugrue, L.P., Corrado, G.S., Newsome, W.T.: Choosing the greater of two goods: neural currencies for valuation and decision making. Nat. Rev. Neurosci. 6, 363–375 (2005)
20. Shors, T.J.: Memory traces of trace memories: neurogenesis, synaptogenesis and awareness. Trends in Neurosciences 27, 250–256 (2004)
21. Treur, J., Umair, M.: On Rationality of Decision Models Incorporating Emotion-Related Valuing and Hebbian Learning. In: Lu, B.-L., Zhang, L., Kwok, J. (eds.) ICONIP 2011, Part III. LNCS (LNAI), vol. 7064, pp. 217–229. Springer, Heidelberg (2011)
22. Weinberger, N.M.: Specific Long-Term Memory Traces in Primary Auditory Cortex. Nature Reviews: Neuroscience 5, 279–290 (2004)

Negotiation Game for Improving Decision Making in Self-managing Teams

M. Birna van Riemsdijk, Catholijn M. Jonker, Thomas Rens, and Zhiyong Wang

Delft University of Technology, The Netherlands
{m.b.vanriemsdijk,c.m.jonker}@tudelft.nl, rensbos@hotmail.com,
wzy19840102@gmail.com

Abstract. This paper presents a game intended to train teams that have both individual and teams goals to negotiate collaboratively in order to reach the team goal in the best way possible. We consider self-managing teams, i.e., teams that do not have a hierarchical structure. The importance of the team goal in comparison to their individual goals is touched upon, as are various conflicts that can occur during such a negotiation. The game, which is implemented in the Blocks World 4 Teams environment, gives a team a specific scenario and allows them to negotiate a plan of action. This plan of action is then performed by agents, after which the team members will be debriefed on their performance.

1 Introduction

In many situations groups of people need to make a joint decision to reach a common goal, while at the same time all participants also try to achieve their individual goals. Examples are friends going on a holiday or planning an evening downtown, teams of researchers negotiating about what proposal to submit and how to divide the budget, incident management teams, design teams of complex systems such as airplanes, collaborating adventure companies that organize a week programme for their clientele.

Our own motivation originated in our research into group decision making for incident management, in which the fire department, police, ambulances, and others have to work together to handle the incident as well as possible. The expectation was that incidents are managed by teams having a clear leader. However, an evaluation of crises has shown that the crisis management decision making process should be seen as a multi-party multi-issue negotiation, as each of the parties in one of the decision-making teams has their own interests and preferences [7] and the nominal leader cannot change that. What motivated us to develop a training game is that in such cases all participants are people dedicated to handle the incident for the good of society, and still they tend to forget this common goal, as they are focusing on the interests associated with their own specialty. When talking to experts involved in airplane design, the issue came up as well: each discipline expert tends to focus on his own problems, and tends to forget to some extent about the airplane they want to create as a team. Once seeing such a pattern, it can be seen in many aspects of life; at work, at home, with friends.

Common in these situations is that in order to make a decision each party has to make some compromises, and the final decision is based on the attitude, negotiation strategy and negotiation skills of each party. Importantly, the authors pose that the team

W. Ding et al. (Eds.): Modern Advances in Intelligent Systems and Tools, SCI 431, pp. 63–68.

will perform better if each team member takes the preferences of other team members into account. When the group decision is not that good, typically, at the start of the process team members do not have this knowledge, nor try to gain this knowledge during the process. Instead they negotiate competitively, meaning that they will focus on and prioritize their own preferences.

Our working hypothesis is that teams in which the team members have individual goals next to the team goal can be trained to adopt a collaborative instead of a competitive mind-set. In this paper we present the design of an agent-based game to train precisely this.

Our game is played by a team that has one group goal to achieve, with each team member also having his own goal to achieve. A negotiation phase is used to create the plan of action, after which software agents play the game based on the negotiation outcome. This allows the players to see the effect that their negotiation result has on the performance. At the end of such a round a debriefing is given to summarize the results to the team and let the team reflect on their performance.

The paper is structured as follows: Section 2 describes related work and Section 3 details the overall design of the game. Section 4 concludes this paper.

2   Related Work

Negotiation is one of the main procedures for dealing with opposing preferences [2]. Such a negotiation can take several forms, either a bilateral negotiation (between two parties only) or a multi-party negotiation (between more than 2 parties) [3]. The topic of negotiation can usually be divided into issues, where each issue has a set of possible values for that issue. Each actor in the negotiation has a preferred outcome, depending on their personal preferences, which will most often not be shared by the other actor(s) as they have differing interests.

Various games have been created that incorporate teamwork and/or negotiation to some extent. In this section we will highlight the two that come closest to our setting.

The SimParc project [1] focuses on participatory park management. The goal of the game is to make players understand conflicts and make them able to negotiate about them. The game takes place in a park council, which consists of various stakeholders like the community or the tourism operator. The topic of discussion is the zoning of the park which entails the desired level of conservation for each part of the park. Each stakeholder has a different preference concerning the zoning for each part of the park, which quickly leads to conflicts of interest. The park manager role acts as an arbiter and makes the final decision based on the final input from the other players. The players can freely negotiate with one or more others. At the end all players can adjust their proposal before the park manager makes a final decision. Players receive information about their performance and can ask the park manager to explain his final decision. The SimParc game comes close to the types of situation we would like to address as it contains a group goal (zoning the park), and individual goals. However, the park manager makes a final decision. In situations we want to address there is no team leader, instead every team member has equal say.

Colored Trails [6] is game that is also close to what we want. It was developed for testing the decision-making procedures in task-oriented settings; it is used for examining agent and human behavior. It focuses on the interaction between the individual goal and the group goal. The game can be played by more than 2 players and consists of a rectangular board containing colored squares. The players get a starting position on this board, a goal position and an initial set of chips. The objective for the players is to reach the goal square. Players can only move to a square which is adjacent to the one they currently occupy. A move has to be paid for with a chip having the same color as the square the player wants to move to. Players negotiate bilaterally to exchange chips. After the game is over, the results are calculated using a scoring function which uses the number of chips the player has left, the distance to the goal position, the number of moves made by the player and whether or not the player has reached the goal state. Group goals can be modeled by adding a scoring component that can only be maximized when all players reach their goal and are therefore composed of multiple individual goals. However the negotiation protocol of Colored Trails is limited to bilateral negotiations, while we want to support situations that require multi-party negotiations.

3 Game Design

In this section, we outline the design of the game that we want to use for training multi-party negotiation.

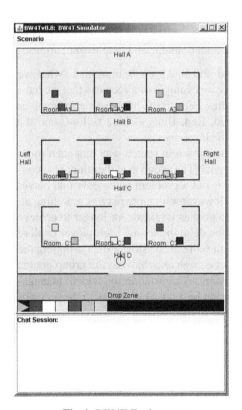

Fig. 1. BW4T Environment

For our game we use the Blocks World For Teams (BW4T) environment as a basis [5]. This environment was originally created to study human-agent teamwork. Our initial reasons for choosing this environment were that it is simple and that it is not directly related to crisis management. The latter is based on the assumption that virtual crisis management environments may not be realistic enough for incident management experts, which might hamper their immersion in the game and reduce the training effect. Furthermore, the game should be effective for any situation in which teams without a clear hierarchy have to make decisions to reach a team goal, while all team members also try to satisfy their individual preferences. An abstract environment, therefore, seems appropriate.

The BW4T environment, shown in Figure 1, consists of 9 rooms in which colored blocks are hidden. One or more simulated robots, which can be controlled by humans or software agents, can traverse these rooms in order to find these blocks.

The goal of the game is to collect certain colors in a preordained order, depicted below the grey area called Drop Zone. Players can pick up one block at a time and can bring it back to and drop it in the Drop Zone. Players can send messages to other players, for example telling them what blocks they have found in a room. None of the players can see the other players.

BW4T as proposed by Johnson et al. includes a group goal, but no individual goals for the players. Therefore, we extend BW4T to BW4T-I by assigning individual goals to each player. That way we can create conflicts between the group goal and individual goals of players, and between the individual goals themselves. Creating these conflicts is necessary as they occur in the situations for which we want to train people. We consider a conflict between two goals to exist when the achievement of one goal hinders the achievement of the other. Various kinds of individual goals may be thought of, creating different kinds of conflicts. Whether a conflict with the group goal occurs, depends also on the choice of the group goal. The criteria we propose for selection of individual goals are as follows:

- High severity of conflict, as a higher amount of conflict should improve awareness of these conflicts in players.
- Possibility of an integrated plan, meaning the possibility of creating a plan that achieves all goals with a lower performance impact. This should show players of our game that by negotiating collaboratively these solutions can be found.
- Ideally there is a way to translate the sets of individual goals and the joint goal to particular problems in the incident management domain, for comparison purposes.

A combination of individual goals, group goal and block configuration, is what we define as a scenario for the game. Below we give an example of a scenario that we have been implemented in our game, and that satisfies our criteria. This scenario involves the group goal "Red, Blue, Green, Yellow, Red, Red, Blue, Green, Yellow" and the individual goal (for all team members) to search all rooms.

The block configuration for this scenario was chosen in such a way that each room has to be explored in order to achieve the group goal. The group goal therefore also contains nine colors as there are nine rooms. This is not a problem when only one player plays this scenario and searches all the rooms. However with more players searching all the rooms, this creates a conflict with the group goal as it takes a lot longer to achieve it. An integrated plan can be created to lessen the delay by letting each player explore the rooms in a different order; however the result will still be slower than dropping the individual goals entirely. The amount of conflict is high as achieving the group goal is delayed greatly by all players searching every room. A translation to incident management could be the fire brigade that would try to get best possible access to the scene of the incident, whereas the ambulances would like to use the same resources (roads) to transport the wounded, and the police needs them for evacuating the population. The number of routes in and out of a quarter of city are limited and should be divided in such a way that the individual goals of all partners are respected as much as possible, while providing the best possible way of handling the incident.

Our game starts with a negotiation phase in which the team members make a joint decision on how they are going to play the game, depending on the group and individual goals. They need to reach an outcome in ten minutes, which emulates the time pressure in the crisis management decision making process. This outcome is given to the agents that play the game using the agreements in the outcome. This allows the team members to see the direct effect of their negotiation outcome, and also prevents team members from changing their plans during the game. After the agents finish the game, a debriefing is given to the team members. The debriefing phase summarizes the results of the negotiation, including how the team members behaved during the negotiation, as well as the resulting performance of the agents.The next round starts after this debriefing. It was necessary to create a fixed negotiation domain for the game, as the agents need to be able to work with the negotiation outcome. In order to create this domain, we conducted experiments with the purpose of determining the issues and values in this domain. In the experiment, participants were allowed to freely negotiate, and in this way relevant issues and values were determined.

As the agents should show the effect of a certain negotiation outcome all of their reasoning uses this outcome. The agents follow a task structure containing three tasks. These tasks are the exploration, delivery and drop off tasks. The exploration task is performed first. This task is used to find the next block that the agent should collect. The delivery task is used to pick up the next block relevant for achieving the team goal. The agent will continue exploring if the next color that the agent should pick up is not the next color in the team goal list. The drop off task is used for dropping a block in the drop zone.

4   Conclusion

This paper presents a training game for training people to negotiate collaboratively in self-managing teams in which there are group goals as well as individual goals in order to create conflicts. Our game design extends the BW4T environment with individual goals. BW4T already has a group goal. Per scenario, team members are allowed 10 minutes to negotiate their joint plan of action. Software agents were implemented to play the game based on the negotiation outcome. The team members are debriefed using the performance of the agents and the results of the negotiation. This should lead to a change in negotiation behavior in the next round.

In the future we intend to perform an experiment in order to test whether our devised game actually changes the perception towards individual and group goals in participants. Moreover, we will investigate the use of automated negotiation agents, using the Genius environment [4], to replace part of the human participants in the first phase of the game. These agents should be endowed with the ability to exhibit different negotiation styles, corresponding with more or less collaborative behavior.

References

1. Briot, J.-P., Sordoni, A., Vasconcelos, E., de Azevedo Irving, M., Melo, G., Sebba-Patto, V., Alvarez, I.: Design of a Decision Maker Agent for a Distributed Role Playing Game – Experience of the SimParc Project. In: Dignum, F., Bradshaw, J., Silverman, B., van Doesburg, W. (eds.) Agents for Games and Simulations. LNCS, vol. 5920, pp. 119–134. Springer, Heidelberg (2009)

2. Carnevale, P.J., Pruitt, D.G.: Negotiation and mediation. Annual Review of Psychology 43, 531–582 (1992)
3. Crump, L.: Multiparty negotiation: what is it? ADR Bulletin 8(7) (2006)
4. Hindriks, K.V., Jonker, C.M., Kraus, S., Lin, R., Tykhonov, D.: Genius: negotiation environment for heterogeneous agents. In: Proceedings of the Eighth International Joint Conference on Autonomous Agents and Multiagent Systems (AAMAS 2009), pp. 1397–1398 (2009)
5. Johnson, M., Jonker, C., van Riemsdijk, B., Feltovich, P.J., Bradshaw, J.M.: Joint Activity Testbed: Blocks World for Teams (BW4T). In: Aldewereld, H., Dignum, V., Picard, G. (eds.) ESAW 2009. LNCS (LNAI), vol. 5881, pp. 254–256. Springer, Heidelberg (2009)
6. Moura, A.V.: Cooperative behavior strategies in colored trails. Master's thesis. Department of Computer Science, Harvard College, Cambridge, Massachussets (2003)
7. van Santen, W., Jonker, C.M., Wijngaards, N.: Crisis decision making through a shared integrative negotiation mental model. International Journal of Emergency Management 6, 342–355 (2009)

Data Mining and Intelligent Systems

ProteinNET: A Protein Interaction Network Integration System

Tianyu Cao[1], Xindong Wu[1], and Xiaohua Hu[2]

[1] Department of Computer Science, University of Vermont, USA
[2] College of Information Science and Technology, Drexel University, USA

Abstract. Protein interaction networks are of great importance to understand the biological cellular process. Recent advances in high-throughput detection methods, such as the yeast two hybrid method, have given researchers access to large volumes of PPI network data. However, protein interaction databases published by different research groups use different protein naming systems and have different levels of reliability. In this paper we design ProteinNET, a protein interaction network integration system, to integrate protein interaction databases from different sources and improve the quality of the protein interaction network by using noise reduction techniques. In addition, the proteinNET system provides five methods to visualize the protein interaction network. The ProteinNET system can help researchers explore the protein interaction network of different data sources.

Keywords: Protein Interaction Networks, Integration, Visualization, Noise Reduction.

1 Introduction

Recent progress in mass spectrometry, two-hybrid methods, tandem affinity purification(TAP) and other high throughput methods has resulted in rapid growth of data that provides a systematic view of protein interactions[12]. Protein interactions give researchers insight into the biological cellular process. In this paper, we aim to build a system that is capable of integrates protein interaction network data from different data sources and improve the reliability of the integrated network by reducing noise from the networks. There are many general graph integration and visualization software packages available such as Cytoscape[11] and Gephi[3]. [10] gives a thorough review of the advantages and disadvantages of these tools. To the best of our knowledge, our system is the first network integration system to incorporate noise reduction techniques. The second advantage of our system is that it is a lightweight system compared to the above systems. It is implemented as a Java applet. Therefore any system with a browser and JVM can run it.

The paper is organized as follows. Section 2 illustrates the architecture of the system. Section 3 introduces the integration module of the system. It discusses two issues and our solution. After that, Section 4 introduces five graph visualization algorithm and shows the visualization results. Section 5 reviews well known noise reduction measures such as the Czekanowski-CDice distance, and the function similarity weight. We also

W. Ding et al. (Eds.): Modern Advances in Intelligent Systems and Tools, SCI 431, pp. 71–76.
springerlink.com © Springer-Verlag Berlin Heidelberg 2012

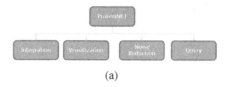

(a)

Fig. 1. System Architecture

compared the performance of the Czekanowski-CDice distance, the function similarity weight and the Katz's measure. Finally, we conclude in Section 6.

2   System Architecture

The ProteinNET system includes four modules. The system architecture is illustrated in Figure 1(a). The integration module resolves differences between multiple data sources; more specifically, multiple data formats and different protein naming systems. The visualization module is responsible for displaying the integrated protein interaction network. The noise reduction module is used to improve the quality of the integrated protein interaction network. The query module gives a simple access point for users. The results of integration, noise reduction and query are all fed to the visualization module directly. The ProteinNET system is focused on improving the quality of the protein interaction network while integrating interaction data from different data sources.

3   The Integration Module

In this section, we discuss two major issues in integration protein interaction networks. The first issue is the differences of data schemas of multiple data sources. To tackle this issue, we narrow our system to three popular data schemas. The second issue is the protein identifier translation problem. We solve this issue by mapping all protein names into UniprotKB access numbers.

To solve this issue, we only select protein interaction databases that conform to the PSI-MITAB format, which was defined by the HUPO Proteomics Standards Initiative. The protein interaction datasets that we collected include DIP, Biogrid, matrixdb, Mint, Intact, MIPS and some other datasets. There are two issues with respect to integration of networks from different data sources. All of these databases supports the PSI-MITAB format.

For the second issue, the system tries to map each identifier to a UniprotKB access number. The most popular identifiers in protein interaction datasets include UniprotKB AC, RefSeq(NCBI Reference Sequences), ordered locus name, gene name, protein name and other data source specific IDs such as Biogrid ID and DIP ID. The UniProt Knowledgebase (UniProtKB) is the central access point for extensive curated protein information, including functions, classifications, and cross-references. It consists of two sections: UniProtKB/Swiss-Prot which is manually annotated and is reviewed and UniProtKB/TrEMBL which is automatically annotated and is not reviewed. The UniprotKB database provides a identifier translation table that can translate gene

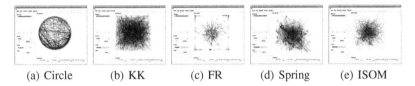

| (a) Circle | (b) KK | (c) FR | (d) Spring | (e) ISOM |

Fig. 2. Five Graph Layouts

names, ordered locus names, Refseq, protein names, DIP ID, MINT ID, gene names and many other identifers to the UniprotKB AC identifiers. Note that Biogrid identifier is not included in this translation table.

In short, in the integration module, we narrow the database schema to three formats, the PSI-MITAB format, the PSI MI XML format and the pajerk network format. We use UniprotKB AC to identify the protein in the protein interaction network. Identifier mappings are performed. For cases where the mappings are not unique, all possible combinations of the uniprotKB AC are included. For cases where there is no mapping between identifier and the UniprotKB AC identifer, we simply drop that interaction.

4   The Visualization Module

In this section we introduce five graph layouts used in our system, including the basic ideas of the algorithms. After that we show images of the graph layouts.

In our system, we have five different layouts of graphs, which are implemented in the package JUNG[1]: the circle Layout, the Fruchterman-Rheingold layout[2], the Kamada-Kawai layout[7], the Spring layout and the ISOM layout[9]. They can be categorized into three categories. The circle layout is the simplest type. The nodes are uniformly distributed on a circle. The second type is the force-based or force-directed algorithms. The Fruchterman-Rheingold layout, the Kamada-Kawai layout, the Spring layout are in this category. These algorithms try to position the nodes of a graph in a two dimensional or three dimensional space so that all the edges are of more or less equal length and there are as few crossing edges as possible. The general idea is to work on a physical model of the graph in which the nodes are represented by steel rings and the edges are springs attached to these rings. If nodes are allowed to move in such a physical model, it will move to an equilibrium state where the potential energy is minimized. The difference between these three layouts is their ways to get to the minimum energy. The ISOM layout belongs to the third category. The ISOM algorithm is based on self-organizing maps. Its main advantage is that it can be very adaptable to arbitrary types of visualization spaces. It can be mapped into both $2D$ and $3D$ visualization spaces. Figure 2 shows all the five layouts.

5   Protein Interaction Network De-noising

In this section, we review three topological based methods of noise reduction. They are the Czekanowski-CDice distance, function similarity weight and Katz's measure. After that we evaluate the performance of these three methods on the yeast interaction networks in terms of function homology and sub cellular location homology. Finally we show the noise reduction system interface.

5.1 Topology Based Noise Reduction Methods

The Czekanowski-CDice distance [4] is based on the idea that if two proteins share some common neighbors, the probability that their interaction is true positive is high. The Czekanowski-CDice distance was firstly devised for the purpose of protein function prediction. It turns out it is very effective in removing the noise of protein interaction networks. It is defined in formula (1),

$$CD - Dist(u, v) = \frac{2|N_{u,v}|}{|N_u| + |N_v|} \qquad (1)$$

where $N_{u,v}$ is the set of common neighbors of u and v, and N_u is the set of neighbors of u and the node u itself.

The function similarity weight [5] is very similar to the Czekanowski-CDice distance. The difference is that the function similarity weight penalizes similarity weights when any of the proteins has too few interacting partners. It is defined in formula (2),

$$Fs - weight(u, v) = \frac{2|N_{u,v}|}{|N_u| + |N_{u,v}| + \lambda_{u,v}} * \frac{2|N_{u,v}|}{|N_v| + |N_{u,v}| + \lambda_{v,u}} \qquad (2)$$

where $\lambda(u, v)$ is used for the purpose of penalizing the proteins with too few interacting partners. It is defined in formula (3),

$$\lambda(u, v) = \max(0, n_{avg} - (|N_u - N_v| + |N_u \cap N_v|)) \qquad (3)$$

where n_{avg} is the average number of level-1 neighbors that each protein has in a dataset.

Katz defined a measure that directly sums over this collection of paths, exponentially damped by length to count short paths more heavily.

$$score(x, y) = \sum_{l=1}^{\infty} \beta^l * |paths_{x,y}^{<l>}| \qquad (4)$$

This measure was shown as a promising measure for the purpose of link prediction in social networks in [8].

5.2 Evaluation of Noise Reduction Methods

In this section, we introduce metrics to evaluate different noise reduction methods. Note that we do not have a label for each protein interaction. Therefore the classification accuracy of a noise reduction method cannot be measured directly. If the noise reduction method can improve the percentage of protein pairs that shares a common cellular function or cellular location, this implies that the noise reduction method improves the quality of the protein interaction dataset. As stated earlier, a large fraction of proteins are not annotated with the information of their cellular location or cellular function.

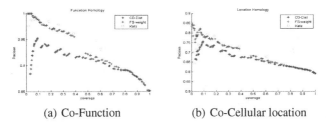

(a) Co-Function   (b) Co-Cellular location

Fig. 3. Evaluation of CD, FSW and Katz's measure

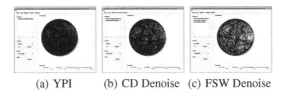

(a) YPI   (b) CD Denoise   (c) FSW Denoise

Fig. 4. Protein Interaction Networks denoised by FS-weight and CD-Dist measure

It is difficult to use this method to evaluate different noise reduction methods on all the datasets. However, the yeast proteins are thoroughly annotated and therefore can be used as a benchmark. We extracted protein function annotation from the MIPS function catalog. We also extracted protein cellular location information from the UniprotKB database. After that, we conducted experiments on the DIP yeast core dataset. We compared Katz's measure against CD-Dist and FS-weight only. This is because it was shown in [6] that FS-weight and CD-dist give better results than the generality index and the IRAP measure. Figure 3 shows the performance of three measures we mentioned in the previous section. Katz's measure and FS-weight are both better than CD-Dist. The horizontal axis denotes the coverage, which means the percentage of pairs of proteins that are included in the whole datasets. The vertical axis indicates the precision, which means the percentage of the pairs of proteins that share a common function(2(a)) or a common cellular location(2(b)). This is like an ROC curve. It can be seen that Katz's measure's performance is similar to FS-weight. They are both better than CD-Dist. However, the points corresponding to the Katz's measure are very scarce compared to FS-weight and CD-Dist. At coverage 0.6, they can achieve more than 90% precision for function homology and 65% precision for location homology.

5.3   The Noise Reduction Module

The noise reduction module is based on the Czekanowski-CDice distance and the function similarity weights. Figure 4 shows the results of noise reduction by the Czekanowski-CDice distance and the function similarity weights. Figure 4(a) is the yeast protein interaction network before denoise action is performed. It can be seen that in Figure 4(b) and Figure 4(c) edges are removed to improve the quality of protein interaction networks.

6 Conclusions

In this paper, we introduced the ProteinNET system, a protein interaction network integration system. The system includes four modules, the integration module, the visualization module, the noise reduction module and the query module. We provided detailed explanations of the four modules. We also presented some in-depth analysis on protein interaction network noise reduction methods and conducted experiments with them. Experimental results have shown that Katz's measure achieves a performance close to state of the art method FS-weight. Our future work will include designing a more sophisticated query module and providing more noise reduction techniques.

Acknowledgments. The research is supported by the US National Science Foundation (NSF) under grants CCF-0905337 and NSF CCF 0905291.

References

[1] JUNG - Java Universal Network/Graph Framework,
    http://jung.sourceforge.net/
[2] Arne Frick, G.S., Wang, K.: Simulating graphs as physical systems. Dr. Dobbs Journal (August 1999)
[3] Bastian, M., Heymann, S., Jacomy, M., Jacomy, M.: Gephi: An open source software for exploring and manipulating networks. In: ICWSM (2009)
[4] Brun, C., Chevenet, F., Martin, D., Wojcik, J., Guénoche, A., Jacq, B.: Functional classification of proteins for the prediction of cellular function from a protein-protein interaction network. Genome Biology 5(1) (2003)
[5] Chua, H.N., Sung, W.K., Wong, L.: Exploiting indirect neighbours and topological weight to predict protein function from protein–protein interactions. Bioinformatics 22, 1623–1630 (2006)
[6] Chua, H.N.N., Wong, L.: Increasing the reliability of protein interactomes. Drug Discovery Today (June 2008)
[7] Kamada, T., Kawai, S.: An algorithm for drawing general undirected graphs. Inf. Process. Lett. 31, 7–15 (1989)
[8] Libennowell, D., Kleinberg, J.: The link-prediction problem for social networks. J. American Society for Information Science and Technology
[9] Meyer, B.: Self-Organizing Graphs - A Neural Network Perspective of Graph Layout. In: Whitesides, S.H. (ed.) GD 1998. LNCS, vol. 1547, pp. 246–262. Springer, Heidelberg (1999)
[10] Pavlopoulos, G., Wegener, A.L., Schneider, R.: A survey of visualization tools for biological network analysis. BioData Mining 1(1), 12+ (2008)
[11] Shannon, P., Markiel, A., Ozier, O., Baliga, N.S., Wang, J.T., Ramage, D., Amin, N., Schwikowski, B., Ideker, T.: Cytoscape: a software environment for integrated models of biomolecular interaction networks. Genome Research 13(11), 2498–2504 (2003)
[12] Uetz, P., Giot, L., Cagney, G., Mansfield, T.A., Judson, R.S., Knight, J.R., Lockshon, D., Narayan, V., Srinivasan, M., Pochart, P., Qureshi-Emili, A., Li, Y., Godwin, B., Conover, D., Kalbfleisch, T., Vijayadamodar, G., Yang, M., Johnston, M., Fields, S., Rothberg, J.M.: A comprehensive analysis of protein-protein interactions in Saccharomyces cerevisiae. Nature 403(6770), 623–627 (2000)

A Parameter Matrix Based Approach to Computing Minimal Hitting Sets

Dong Wang, Wenquan Feng, Jingwen Li, and Meng Zhang

School of Electronics and Information Engineering,
Beijing University of Aeronautics and Astronautics,
Beijing 100191, China
wangdong1106@gmail.com, {buaafwq,lijingwen}@buaa.edu.cn,
zhangmeng501@126.com

Abstract. Computing all minimal hitting sets is one of the key steps in model-based diagnosis. Because of the low capabilities due to the expansion of state space in large-scale system diagnosis, more efficient approximation algorithms are in motivation. A matrix-based minimal hitting set (M-MHS) algorithm is proposed in this paper. A parameter matrix records the relationships between elements and sets and the initial problem is divided into several sub-problems by decomposition. The efficient prune rules avoid the computation of the sub-problems without solutions. Parameterized way and de-parameterized way are both given so that the more suitable algorithm could be chosen according to the cases. The simulation results show that, the proposed algorithm outperforms HSSE and BNB-HSSE in large-scale problems and keeps a relatively stable performance when data changes in different regulations. The algorithm provides a valuable tool for computing hitting sets in model-based diagnosis of large-scale systems.

Keywords: minimal hitting set, model-based diagnosis, parameter matrix.

1 Introduction

The Hitting Set Problem is well known in model-based diagnosis that minimal hitting sets of the set of conflicts are minimal diagnoses that can explain inconsistencies between modeled and observed behaviors [1, 2]. Proven that the hitting set problem is NP-hard, there is a high motivation for more efficient approximation algorithms.

Over the years, a multitude of approaches for computing minimal hitting set have been proposed, The most common is based on HS-trees [3,4,5,6]. Jiang and Lin [7] propose a method using Boolean Formulas. Other approaches using AI technologies such as genetic algorithm and simulated annealing algorithm have been shown in [8, 9]. HSSE [10] solves the problem of losing solutions when the traditional HS-tree is pruned, but its efficiency relies on the disciplinary data. BNB-HSSE [11] is a branch and bound algorithm based on HSSE. It decomposes the given problem into finer and finer partial problems by applying branching operation, but the parameterized way makes the same sub-trees generated over and over again by different branching. The

W. Ding et al. (Eds.): Modern Advances in Intelligent Systems and Tools, SCI 431, pp. 77–85.
springerlink.com   © Springer-Verlag Berlin Heidelberg 2012

algorithm with logic arrays [12] is easier for programming, but it is actually a kind of enumeration which leads to low performance. In [13], the upper bound obtained by mapping the problem into integer programming (LP) optimization is used to eliminate unexpected sub-problems. But the algorithm only finds minimal hitting sets with minimal cardinality. In large scale systems, the size of state space is also a great challenge. Whether an algorithm could solve lager problem than others becomes an important judgment of algorithm.

In this paper, we tackle the problem of finding minimal hitting sets in de-parameterized and parameterized ways. We present an algorithm M-MHS which uses a parameter matrix to describe the relations between elements and subsets and finds solutions by iterations of matrix decomposition. Prune rules terminates the useless iterations as early as possible. To evaluate the overall performance, we compare M-MHS with HSSE and BNB-HSSE by solving different types of problem. The experimental results show that M-MHS could find all expected minimal hitting sets and uses less time than the other two algorithms in most cases, especially in large-scale problems.

The paper is organized as follows. After introducing the relative definitions and corollaries in section 2, we depict the M-MHS algorithm in section 3 and 4, in de-parameterized and parameterized way respectively. Our experimental results and discussions can be found in section 5. Finally, the conclusions are drawn in section 6.

2   Preliminaries

Definition1: Given a set of all elements $U = \cup c_i = \{e_1, e_2, \cdots e_m\}$ and a set of subsets $CS = \{c_i, i = 1, 2, \cdots, n\}$, H is a hitting set if $H \subseteq U$ and $H \cap c_i \neq \emptyset$ for every $c_i \in CS$. H is minimal if no proper subsets of H is a hitting set.

Definition2: The matrix $M_{m \times n}$ describes the attribution of elements to subsets (m is the number of elements in CS, and n is the number of subsets).The $(i, j)^{th}$ entry in the matrix is denoted as m_{ij} and $m_{ij} = 1$ if e_i belongs to c_j otherwise $m_{ij} = 0$.

$$M_{m \times n} = \begin{array}{c} e_1 \\ \vdots \\ e_m \end{array} \begin{bmatrix} \begin{array}{ccc} c_1 & \cdots & c_n \\ m_{11} & \cdots & m_{1n} \\ \vdots & m_{ij} & \vdots \\ m_{m1} & \cdots & m_{mn} \end{array} \end{bmatrix}$$

Corollary 1: If all elements of ith row equal to 1, i.e. $m_{ij} = 1, j = 1, \cdots, n$, then $\{e_i\}$ is a minimal hitting set.

Corollary 2: If all elements of ith row equal to 0, i.e. $m_{ij} = 0, j = 1, \cdots, n$, then e_i does not appear in minimal hitting sets.

Corollary 3: If all elements of jth line equal to 0, i.e. $m_{ij} = 0, i = 1, \cdots, m$, then the problem has no solution; if the matrix has not all-zero line, the problem has at least one solution.

3   M-MHS Algorithmic Approach

3.1   De-parameterized Way

M-MHS counts 0/1 elements in the matrix recording their positions and decomposes the given problem into partial problems. The three corollaries are applied to: 1) find the minimal hitting set of single element rapidly; 2) terminate the partial problems without solution as early as possible. In de-parameterized way, the cardinality of minimal hitting set is not in consideration.

1:    **function** Caculate-HittingSet-DP (m, hs)
2:    **returns** All hitting sets according to matrix
3:    **if** m has all-zero lines
4:    **then** return;
5:    Remove-All-Zero-Row (m);
6:    $order\leftarrow$Count-And-Sort (m);
7:    **loop do**
8:         $row\leftarrow$Front ($order$);
9:         **if** row is an all-one row
10:        **then** add $element$ indicated by row into hs;
11:        **else** $new\_m\leftarrow$Get-Matrix (m, row);
12:         **if** new_m is not empty
13:         **then** Caculate-HittingSet (new_m, new_hs)
14:        Remove-Row(row);
15:        Prune(hs, new_hs);
16:    **end**
17:    **return** hs;

The input m is the matrix after decomposition, and is the initial matrix at the beginning. If the matrix has all-zero lines, the function returns to the upper iteration. Otherwise the sums of all elements in each row of matrix are calculated. After removing the all-zero rows, the left rows are sorted by descending sum. Then the row with largest sum is popped, which the decomposition lies on, i.e. if an element occurs most frequently, it will be considered firstly. Similar to the branch-and-prune algorithm, considering the element e_i to be included or excluded in the hitting set, two matrixes are generated.

M_{out}: Check each entry of the row of e_i. If $e_{ij} = 0$, pick out j^{th} line of matrix and add it into M_{out}.;
M_{in}: Remove i^{th} row from matrix and the left $(m - 1) \times n$ is M_{in};

3.2   Parameterized Way

In parameterized way, we only calculate hitting sets with cardinality d. Thus, evaluating the range of possible cardinality is necessary before matrix decomposition to avoid useless iterations.

```
1:   function Caculate-HittingSet-P(m, hs. d)
2:   returns All hitting sets of cardinality d according to matrix
3:   if m has all zero lines
4:   then return;
5:   Remove-All-Zero-Row (m);
6:   order←Count-And-Sort (m);
7:   loop do
8:          row←Front (order);
9:          Evaluate lower and upper bounders;
10:         if d is not in the interzone
11:         then return;
12:         if row is an all-one-row and d equals to 1
13:         then add element indicated by row into hs;
14:         else if row is not an all-one-row
15:         then new_m←Get-Matrix (m, row);
16:          if new_m is not empty
17:          then Caculate-HittingSet (new_m, new_hs, d-1)
18:         Remove-Row(row);
19:         Prune(hs, new_hs);
20:  end
21:  return hs;
```

When the next expanding element is popped from the order list, the lower and upper bounds are evaluated first:

$$B_l = 1 + \operatorname{ceil}\left(\frac{matrix.\mathrm{col\_size} - row.\mathrm{cnt1}}{row.\mathrm{cnt1}}\right)$$

$$B_u = 1 + \left(matrix.\mathrm{col\_size} - row.\mathrm{cnt1}\right) \geq \left(matrix.\mathrm{row\_size} - 1\right)?$$
$$\left(matrix.\mathrm{col\_size} - row.\mathrm{cnt1}\right) : \left(matrix.\mathrm{row\_size} - 1\right)$$

where row.cnt1 is the number of 1-entries in row. If d is in the range, the matrix will be decomposed next; otherwise, the function returns. Note that the parameterized way has special requirement on the cardinality. Thus, hitting sets with single element are added into hittingsets list iff the current d equals to 1.

3.3   Prune Rules

There are three rules in function Prune(hs, new_hs);:

1)   For a hitting set s of new_hs, if there is a subset of s in new_hs, delete s;
2)   For a hitting set s of new_hs, if there is a superset of s in new_hs, replace the superset with s;

3) If *new_hs* is empty, then terminate this iteration and jump to the next. In this partial problem, all the elements with less occurrence than the current element cannot be a member of hitting set;

The first two rules are obvious and used widely in most algorithms. The third rule is derived from the 3rd corollary.

3.4   Example

Given a set $CS = \{\{B, C\}, \{A, C\}, \{A, D\}, \{B, D\}, \{A, B\}\}$. The initial matrix is:

$$
M = \begin{array}{c c c c c c c}
 & c_1 & c_2 & c_3 & c_4 & c_5 & \\
A & 0 & 1 & 1 & 0 & 1 & 3 \\
B & 1 & 0 & 0 & 1 & 1 & 3 \\
C & 1 & 1 & 0 & 0 & 0 & 2 \\
D & 0 & 0 & 1 & 1 & 0 & 2
\end{array}
$$

The last column shows the sums of all elements in each row.

1) In de-parameterized way
A and B are both the maximum. Suppose we choose A as the expanding element. If A is included in hitting set, the sub matrix after removing all-zero rows is:

$$
M_A = \begin{array}{c c c c}
B & 1 & 1 & 2 \\
C & 1 & 0 & 1 \\
D & 0 & 1 & 1
\end{array}
$$

Because the row of B is an all one row, B is a hitting set of this partial problem. Add B into the hitting set list $HS_A = \{\{B\}\}$ and remove the row of B. Choose C as the next expanding element. Considering C in the hitting set, the sub matrix is:

$$M_{AC} = D \ \ 1 \ \ 1$$

Thus D is a hitting set. Combining D with the expanding element C, $HS_A = \{\{B\}, \{C, D\}\}$. Removing the row of C from M_A, the matrix has only one row left and the first entry is zero. With the third prune rule, the iteration of A ends and $HS = \{\{A, B\}, \{A, C, D\}\}$. A is removed from M and B is picked. Likewise, we can conclude that $HS_B = \{\{C, D\}\}$. Further, the hitting set list is updated by $HS = \{\{A, B\}, \{A, C, D\}, \{B, C, D\}\}$.

Then C and D are left in M. Notice that the 5th line is an all-zero line. Thus the problem has no solution anymore and the iteration of the top level ends returning $HS = \{\{A, B\}, \{A, C, D\}, \{B, C, D\}\}$.

2)   In parameterized way

Suppose the initial $d=2$ and we choose A as the expanding element. The lower bound is 1 and the upper bound is 4. As $1<d<4$, the sub matrix is:

$$M_A = \begin{array}{c} B \\ C \\ D \end{array} \begin{array}{ccc} 1 & 1 & 2 \\ 1 & 0 & 1 \\ 0 & 1 & 1 \end{array}$$

Because the row of B is an all-one row and now $d=1$, B is added into the hitting set list $HS_A = \{\{B\}\}$, then remove the row of B. For C or D, the lower bound is 2, so there is no other hitting set of cardinality of 1 in this partial problem. Finally, only $\{A, B\}$ survives.

4   Experimental Results

4.1   De-parameterized Emulation

To evaluate the overall performance, we compare M-MHS with HSSE and BNB-HSSE. The original BNB-HSSE is modified to make the algorithm de-parameterized. The first modification is removing the calculation of upper and lower bounds; the second is rewriting the condition of node branching. The code of M-MHS is written in C++ and all experimental results are measured on a 2.66 GHz Intel Core2 Duo CPU PC with 1.95GB of memory.

1)   Random Data Experiments

Table 1 shows the average time used by these algorithms on 50 random problems. In HSSE, the enumeration is restricted within 6 elements, i.e. the hitting sets with cardinality less than 7 are calculated. The time of enumerate all minimal hitting sets is far more than that. Modified BNB-HSSE outperforms M-MHS in small problems, but in large problems M-MHS is more capable.

Table 1. Average performance on 50 random problems (*enumeration incomplete)

Element number	Set number	Running time(ms)		
		HSSE	Modified BNB-HSSE	M-MHS
25	20	1762*	146.9	264.2
26	23	2963*	163.3	351.6
30	25	5586*	1153	989.5
31	28	7559*	2565	1983
35	30	15266*	14142	7178
36	33	21418*	10078	8235
40	35	40133*	150655	90956
41	38	64484*	235253	80971

2)   Data of Different Regulations Experiments

The number of sets is n and the elements are integers from 0 to n. The length of the sets depends on the parameter i, which impacts the cardinality of hitting set indirectly. The kth set contains $n - i$ continuous elements starting from k, i.e.

$$\begin{cases} \text{if } k \leq i+1 \quad \{k, k+1, \dots n-i+k-1\} \\ \text{else } \{k, k+1, \dots n, 0, \dots k-i-2\} \end{cases}$$

Notice that the second experiment is a special case of $i = 1$. Consider $n = 50$ and $i \in [1, 10]$.

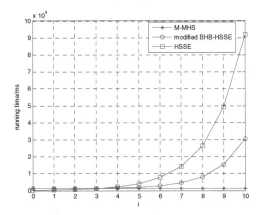

Fig. 1. Running times of different fixed data

Along with the increasing of i, the performances of HSSE and modified BNB-HSSE decrease apparently, while M-MHS stays in a steady capability. In BNB-HSSE, the lager i is, the less an element appears. Therefore, the branch "sets without the specified element" of the expanding tree grows exponentially, which impacts the solving time. But for M-MHS, all the initial matrixes have the same size and i only impacts the number of lines of the each M_{out}, which means that the times of decomposition are almost the same.

4.2   Parameterized Emulation

In parameterized way, we evaluate the performance of M-MHS and original BNB-HSSE on 50 random problems, which are identical with the de-parameterized experiment. Table 2 shows the results when the cardinality of hitting sets d changes from 2 to 5. The average performance of M-MHS is better than that of BNB-HSSE in all cases. The disparity is more obvious when d is large or the problem is massive.

Table 2. Average performance on 50 random matrixes with different d

Element number	Set number	Running time (ms)							
		d=2		d=3		d=4		d=5	
		M-MHS	BNB-HSSE	M-MHS	BNB-HSSE	M-MHS	BNB-HSSE	M-MHS	BNB-HSSE
25	20	58.13	127.3	92.27	178.7	158.5	226.2	241.7	289.6
26	23	68.73	157.4	113.9	215.2	188.1	262.3	276.1	325.8
30	25	109.9	261.2	175.5	369.0	298.7	462.7	454.6	614.5
31	28	142.1	357.8	211.6	554.9	377.1	742.0	649.0	1011
35	30	198.0	592.0	302.3	1111	578.9	1667	1176	2670
36	33	236.4	803.9	385.5	1478	772.4	2089	1513	3081
40	35	318.3	1302	524.4	3185	1123	6226	2776	11590
41	38	384.1	1477	617.4	3916	1323	8091	3849	16426

5   Conclusion

In this paper, we present a matrix based approach to computing minimal hitting sets, in both de-parameterized way (all hitting sets) and parameterized way (d-hitting sets). The problem is mapped onto a matrix and the branch and prune rules are related to the judgment of 1/0 entries. The experiment results show that in large scale problems, M-MHS has better overall performances than HSSE and BNB-HSSE. But for the problems of small scale or of special regulation, M-MHS couldn't find solutions faster than the others because of matrix computation. Thus this method provides a valuable tool for solving large scale problems.

References

1. de Kleer, J., Williams, B.C.: Diagnosing multiple faults. Artificial Intelligence 32(1), 97–130 (1987)
2. Williams, B.C., Ragno, R.J.: Conflict-directed A* and its role in model-based embedded systems. Discrete Applied Mathematics 155(12), 1562–1595 (2007)
3. Reiter, R.: A theory of diagnosis from first principles. Artificial Intelligence 32(1), 57–95 (1987)
4. Greiner, R., Smith, B.A., Wilkerson, R.W.: A correction to the algorithm in reiter's theory of diagnosis. Artificial Intelligence 41(1), 79–88 (1989)
5. Wotawa, F.: A variant of Reiter's hitting-set algorithm. Information Processing Letters 79(1), 45–51 (2001)
6. Yunfei, J., Li, L.: Computing the Minimal Hitting Sets with Binary HS-Tree. Journal of Software 13(12), 2267–2274 (2002) (in Chinese)
7. Yunfei, J., Li, L.: The Computing of Hitting Sets with Boolean Formulas. Chinese Journal of Computers 26(8), 919–924 (2003) (in Chinese)
8. Jie, H., Lin, C., Peng, Z.: A Compounded Genetic and Simulated Annealing Algorithm for Computing Minimal Diagnosis. Journal of Software 15(9), 1345–1350 (2004) (in Chinese)
9. Li, L., Jiang, Y.: Computing minimal hitting sets with genetic algorithm. In: Proceedings of the 13th International Workshop on Principles of Diagnosis, Austria, pp. 77–80 (2002)

10. Zhao, X., Ouyang, D.: A method of combining SE-tree to compute all minimal hitting sets. Progress in Natural Science 16(2), 169–174 (2006)
11. Chen, X., Meng, X., Qiao, R.: Method of computing all minimal hitting set based on BNB-HSSE. Chinese Journal of Scientific Instrument 31(1), 61–67 (2010) (in Chinese)
12. Li, L.: Computing minimal hitting sets with logic array in model-based diagnosis. Journal of Jinan University (Natural Science) 23(1), 24–27 (2002) (in Chinese)
13. Fijany, A., Vatan, F.: New Approaches for Efficient Solution of Hitting Set Problem. In: ACM International Conference Proceeding Series (2008)

A Scalable Feature Selection Method to Improve the Analysis of Microarrays*

Aida de Haro-García, Javier Pérez-Rodríguez, and Nicolás García-Pedrajas

Department of Computing and Numerical Analysis, University of Córdoba, Spain
adeharo@uco.es, javier@cibrg.org, npedrajas@uco.es
http://www.cibrg.org/

Abstract. DNA microarray experiments are used to collect information from tissue and cell samples regarding gene expression differences that are useful for diagnosis and treatment of many different diseases. The predictive accuracy is hindered by the large dimensionality of these datasets and the existence of irrelevant and redundant features. The performance of a feature selection process could improve the classification accuracy of this demanding research field.

However, standard feature selection method performance may be very poor in high-dimensional microarray data. We propose a scalable evolutionary method to select relevant genes. We use a divide-and-conquer approach to deal with the scalability issues of the evolutionary algorithms, and a combination of different rounds of feature selection to increase the accuracy results and storage reduction. Our proposal improves the results of standard classifiers and feature selection methods in accuracy and storage reduction for 8 different microarray datasets.

1 Introduction

Traditionally, the bottleneck preventing the development of more accurate systems via machine learning was the limited data available. However, recently [2], the limiting factor is learners' inability to use all the data in the available time. This situation is the result of the improved ability to gather information on a massive scale together with the application of machine learning techniques, and more specifically data mining algorithms, to new scientific problems.

Many of the widely used algorithms in machine learning were developed when the typical dataset sizes were much smaller than now. Many of those algorithms are not able to deal with such large amounts of information, being impracticable due to their memory and time requirements or producing low quality results. Thereby, with the growing size of the datasets in all the fields of application of machine learning, the need to scale up data mining algorithms has also increased.

A good example of new demanding fields is bioinformatics and especifically microarray datasets which are characterized by their large dimensionality (up

---

* This work has been financed in part by the Excellence in Research Projects P07-TIC-2682 and P09-TIC-4623 of the Junta de Andalucía.

W. Ding et al. (Eds.): Modern Advances in Intelligent Systems and Tools, SCI 431, pp. 87–92.
springerlink.com   © Springer-Verlag Berlin Heidelberg 2012

to several tens of thousands of genes) and their small sample sizes [10]. In this case, the scalability problem is not about resources or time, but about poor performance. Consequently, feature selection algorithms can play a crucial role to deal with these particular characteristics of microarray data.

Feature selection has been a fertile field of research [7,3],and can be defined as the selection of a subset of M features from a set of N features, $M < N$, such that the chosen subset optimizes the value of some criterion function over all subsets of size M [8].

The advantages of feature selection come at a certain price, as the search for a subset of relevant features introduces an additional layer of complexity to the modelling task. This new layer increases the memory and running time require-ments, making these algorithms very inefficient and inaccurate when applied to problems that involve very large datasets.

In order to be able to efficiently work with large dimensionality microarray datasets and obtain precise classifications, we propose a scalable methodology that performs several rounds of evolutionary fast feature selectors.

This paper is organized as follows: Section 2 presents our proposal; Section 3 shows the experimental results; and Section 4 shows the conclusions of our work.

2   A Scalable Evolutionary Feature Selection Method

Our main objective is the application of a feature selection algorithm whose selection of relevant genes may be able to improve the classification results in the complex field of DNA microarrays.The most accurate feature selection methods are the wrapper ones. The inclusion of the classification model within the subset search deals better with feature dependencies in complex datasets. Therefore, we decided to implement a simple genetic algorithm due to the fact that previous work [9] successfully compared a genetic algorithm (GA) approach with classical sequential search (forward and backward), and with a nonoptimal variation of branch and bound which is able to work with a nonmonotonic criterion.

To address the microarray problem, we designed a simple GA algorithm that represents a feature subset as a binary string (a *chromosome*) of length n. A zero or a one in the position i of the chromosome denotes the absence or presence of the feature i in the set. Note that n is the total number of available features. We apply standard genetic operators such as two-point crossover and mutation. At the beginning of each generation we perform an elitism step, and each individual is evaluated by means of the fitness function:

$$fitness(ind) = suc\_rate(i) \cdot \alpha + (1 - \alpha) \cdot ((1 - n\_sel\_feat)/n), \qquad (1)$$

where suc_rate is the wrapper evaluation of the subspace using a given classifier, α is a value in the interval $[0, 1]$ which is set to 0.75, n_sel_feat is the number of selected features.

However, evolutionary algorithms are well-known by their scalability issues when the number of variables is considerably large.

To overcome this problem we propose a new scalable method to efficiently apply our simple genetic algorithm to high-dimensional datasets. Our method is based on a divide-and-conquer approach. The basic step of the method is similar to the stratification approach [1] but selecting features instead of instances. The feature selection algorithm is applied several times to subsets of features, and the results are combined through a voting process. Therefore, our methodology is based on repeating several rounds of fast evolutionary feature selectors. Each round on its own would not be able to achieve a good performance. However, the combination of several rounds using a voting scheme is able to improve the performance of a feature selection algorithm applied to the whole dataset. A method sharing this philosophy has been successfully used for scaling feature selection algorithm to datasets of many instances [6].

More specifically, for each step, i and j number of subsets, we divide the dataset S into several small disjoint subsets of features $S_{i,j}$. It is noteworthy that this partition is performed randomly and that the sum of all subsets in each round results in the whole dataset. After all the n_s rounds are applied, and that can be done in parallel as all of them are independent from each other, the combination method constructs the final selection of relevant features, C, as the result of the feature selection process.

Each application of the feature selection algorithm to a subset contributes with one vote. The voting process is carried out as follows: Each time an feature selection algorithm is applied to a subset, it outputs a selection of features. These features, considered by the algorithm relevant, receive a vote, which is recorded for all features in all subsets and in all rounds. After the r rounds, each feature has received a number of votes in the interval $[0, r]$.

The final combination of votes must set a threshold to decide whether a feature must be selected as the final output of the process. As the performance of this fixed threshold heavily depends on the problem, we developed a method to automatically estimate the optimum threshold for each dataset.

For that aim, we define a "fitness" function that evaluates the goodness of a certain threshold t. Using the threshold t, we obtain the set of selected features, $S(t)$, and then the fitness function, $f(t)$, is calculated as follows:

$$f(t) = \beta r(t) + (1 - \beta)a(t), \tag{2}$$

where $r(t)$ is the storage reduction achieved using threshold t, and $a(t)$ is the accuracy achieved with the features in $S(t)$ using a 1-nearest neighbor (1-NN) classifier over the training set. To obtain the best threshold, all the values in $[0, r]$ are evaluated and the optimum is chosen. $\beta = 0.75$ was used in our experiments.

3 Experimental Results

We have used a set of eight different microarray problems to test the performance of the proposed method, which are shown in Table 1. These high-dimensional biomedical datasets are from the Kent Ridge Bio-medical Repository[1]. We chose

---

[1] http://datam.i2r.a-star.edu.sg/datasets/krbd/.

Kent Ridge datasets because they are highly challenging problems of extended use within the data mining and microarray community.

For estimating the storage reduction and generalization error we used 10-fold cross-validation. The source code, in C and licensed under the GNU General Public License, used for all methods as well as the partitions of the datasets are freely available upon request to the authors. We have used the Wilcoxon test [11] as the main statistical test for comparing pairs of algorithms.

The number of generations in the GA used in our simulation study is set at 100. The number of individuals in the population is set at 100. At the beginning of each generation we apply a 10% of elitism and afterwards we get the rest of the population by means of applying iteratively a two-point-crossover operator (in which we keep the two best individuals of each crossover step). The standard mutation percentage is fixed to 10%. The parameters chosen for the genetic algorithm are fairly standard values [1].

Our scaling method performs $r = 10$ iterations of the evolutionary algorithm. Our experiments showed that more rounds added little improvement. Each round performs a different partition of the dataset into disjoint subsets of $M = 50$ features. As explained in the previous section, β is set to 0.75, because for us it is more important a lesser error than little storage requirements when directly selecting a subset of features.

In the experiments 1, our first aim was to keep less but more informative features does not decrease the accuracy using the whole dataset, with the advantage of requiring less storage resources. In the following stage we compared the results of the standard genetic algorithm using 1-NN with the performance of our scalable version of this same feature selection algorithm using its same parameters. The repetition of the feature selection process of our approach ensures that we have gathered enough information about the genes for the combination step to be useful and consequently provides a better selection of the relevant genes.

Table 1. Accuracy and reduction of the standard genetic algorithm and our approach

Data set	Cases	No.feats	1-NN	Standard GA Reduction	Standard GA Accuracy	Scalable GA (50 feat.) Reduction	Scalable GA (50 feat.) Accuracy
1 all-aml	72	7,129	0.8595	0.6935	0.8595	0.9975	0.9458
2 breast-can.	97	24,481	0.5557	0.6746	0.6157	0.9991	0.6370
3 central-nerv.	60	7,129	0.5462	0.6872	0.6452	0.9995	0.5886
4 colon-tumor	62	2,000	0.7286	0.7055	0.6929	0.9908	0.7071
5 leukemia	72	12,582	0.8417	0.6825	0.8440	0.9981	0.9381
6 ovarian-canc.	253	15,154	0.9366	0.6925	0.9206	0.9955	0.9922
7 pediatric-leuk.	327	12,558	0.7711	0.6716	0.7560	0.9807	0.8815
8 prostate	136	12,600	0.8242	0.6815	0.8313	0.9932	0.9198
Average	134.88	11704.13	0.7579	0.6861	0.7707	0.9943	0.8263

The figure 1 show the usefulness of our approach. In terms of accuracy, our approach obtained better results than 1-NN and the genetic algorithm. The Wilcoxon test found the differences significant at a confidence level of 95%. Besides, the good accuracy was obtained with a reduction on average above the 99%. In terms of reduction, our approach also improved significantly the results of the standard GA according to the Wilcoxon test.

It may be argued that the proposed approach improved the accuracy and reduction of the standard genetic algorithm just because that method is performing poorly. To account for that fact we performed another experiment using SVM-RFE method [5]. This algorithm conducts feature selection in a sequential backward elimination manner, which starts with all the features and discards one feature at a time. The squared coefficients: $w_j^2(j = 1, ...; n\_feat)$ of the weight vector w are employed as feature ranking criteria. Intuitively, those features with the largest weights are the most informative. In an iterative procedure of SVM-RFE one trains the SVM classifier, computes the ranking criteria w_j^2 for all features, and discards the feature with the smallest ranking criterion. One can consider that the removed variable is the one which has the least influence on the weight vector norm. This method is used because it has been proved to be one of the best performing methods for microarray data [4].

Because of the procedure provides as output a ranking of the features, we have compared the results obtained with our method and with SVM-RFE when the same number of features as in our method is selected. These results are shown in Figure 1. The figures shows that our approach obtained better results than SVM-RFE. The differences are significant according to the Wilcoxon test.

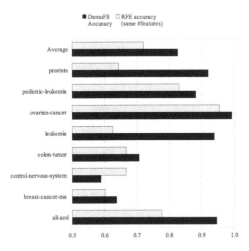

Fig. 1. Accuracy for SVM-RFE and our scalable version with the same number of features selected

4   Conclusions and Future Work

In this paper we have addressed complex microarray classification problems by means of feature selection. Providing the biology researcher a subset of relevant genes allows him/her to focus attention on a manageable subset of promising gene candidates that actively contribute to classification of cell-lines.

We have proved that applying an evolutionary algorithm which uses a kNN evaluator to select the most relevant genes additionally improves the classification accuracy of a regular classifier. We have designed a new method to scale the designed genetic algorithm to overcome the scaling issues that arise when dealing with large dimensionality datasets, as it is our case. The experiments have shown that our scaling approach significantly helps to increase the classification accuracy as well as it decreases the storage requirements.

References

1. Cano, J.R., Herrera, F., Lozano, M.: Using Evolutionary Algorithms as Instance Selection for Data Reduction in KDD: An Experimental Study. IEEE Transactions on Evolutionary Computation 7(6), 561–575 (2003)
2. Craven, M., DiPasquoa, D., Freitagb, D., McCalluma, A., Mitchella, T., Nigama, K., Slatterya, S.: Learning to construct knowledge bases from the World Wide Web. Artificial Intelligence 118(1-2), 69–113 (2000)
3. Dash, M., Choi, K., Scheuermann, P., Liu, H.: Feature Selection for Clustering - A Filter Solution. In: Proceedings of the Second International Conference on Data Mining, pp. 115–122 (2002)
4. Ding, Y., Wilkins, D.: Improving the performance of svm-rfe to select genes in microarray data. BMC Bioinformatics 7(suppl. 2), S12 (2006)
5. Guyon, I., Weston, J., Barnhill, S., Vapnik, V.: Gene selection for cancer classification using support vector machines. Machine Learning 46, 389–422 (2002)
6. de Haro-García, A., García-Pedrajas, N.: Scaling up feature selection by means of democratization. In: García-Pedrajas, N., Herrera, F., Fyfe, C., Benítez, J.M., Ali, M. (eds.) IEA/AIE 2010. LNCS, vol. 6097, pp. 662–672. Springer, Heidelberg (2010), http://portal.acm.org/citation.cfm?id=1945847.1945926
7. Kim, Y., Street, W.N., Menczer, F.: Feature selection in unsupervised learning via evolutionary search. In: The 6th ACM SIGKDD International Conference on Knowledge Discovery and Data Mining, pp. 365–369. ACM Press (2000)
8. Narendra, P.M., Fukunaga, K.: Branch, and bound algorithm for feature subset selection. IEEE Transactions Computer C-26(9), 917–922 (1977)
9. Siedlecki, W., Sklansky, J.: A note on genetic algorithms for large-scale feature selection. Pattern Recognition Lett. 10, 335–347 (1989)
10. Somorjai, R.L., Dolenko, B., Baumgartner, R.: Class prediction and discovery using gene microarray and proteomics mass spectroscopy data: curses, caveats, cautions (2003)
11. Wilcoxon, F.: Individual comparisons by ranking methods. Biometrics 1, 80–83 (1945)

A Methodology for Supporting Lean Healthcare

Daniela Chiocca[1], Guido Guizzi[1], Teresa Murino[1],
Roberto Revetria[2], and Elpidio Romano[1]

[1] Dipartimento di Ingegneria dei Materiali e della Produzione, Univertisty of Naples "Federico II", Naples, Italy
daniela.chiocca@unina.it, g.guizzi@unina.it, murino@unina.it, elromano@unina.it
[2] Dipartimento di Ingegneria Meccanica, University of Genoa, Genoa, Italy
roberto.revetria@unige.it

Abstract. This article focused on application of Lean techniques in healthcare service. In this context, the aim of this work is to demonstrate that the problems associated with the application in the world services of the rules governing production processes are only apparent. In fact, the provision of a service search, like industrial production, to implement value-added processes and also an appropriate use of resources consumption, although until now it was considered that the service management should use management strategies completely different from those used in the industrial sector. The service sector is backward compared to the manufacturing industry in terms of efficiency, automation, repeatability and quality. The reason is very simple: the design and implementation of the process of service delivery have always been much less expensive than the corresponding activities of the manufacturing sector. In particular, we will analyse as case study, vaccination service in the City District in the South of Italy. It will, therefore, applied the techniques of "lean" philosophy to a specific activity of this service in order to improve performance. In particular we are going to focus on the processes of vaccines delivery among the various activities within the District. We have used a new study approach based on simulation : using a software (Powersim), we have developed a model of the current state and, basing on suggestions for improvement; then, after this research, have noticed an improvement of critical performance parameters developing a model of the future state.

Keywords: Health care service management, Lean Technics, System Dynamics, Value Stream Mapping.

1 Introduction

Although you can say that there are service companies since the beginning of civilization, today we continue to experience a trend that leads us towards an increasingly service oriented. To realize this vision of Operational Excellence is working to link the internal flow of the system to the flow of materials from external suppliers, not forgetting the flow of information. Essential to the success of the Lean Supply Chain

W. Ding et al. (Eds.): Modern Advances in Intelligent Systems and Tools, SCI 431, pp. 93–99.
springerlink.com
© Springer-Verlag Berlin Heidelberg 2012

is the mutual benefit of the supplier and the customer. To establish the relationship according to specific procedures, is essential to have well-defined strategy of buy, so that we can support activities that will be implemented to establish the Lean Supply Chain. To this end, the application of Lean philosophy to the supply chain is the main instrument for delivering an efficient and effective. On this trail, after outlining the principles that underlie the paradigm of Lean and Supply Chain, the aim of present study is to analyze how and to what extent, service firms can reap the benefits of the Lean Supply Chain Management, borrowing, thus, from the manufacturing approaches that until now was considered to be the sole prerogative of that context. .The purpose of this approach is to eliminate the causes of waste in a process , with the subsequent optimization of production flows. First we describe the origins of lean production, known as the first industrial application of lean thinking, the main interventions carried for the implementation of this philosophy and development over the years. In the second section we will describe the key concepts of lean thinking, including those of value and waste, and the characteristic tools to develop a lean company. Finally, we will describe the evolution of the lean philosophy from its application in manufacturing up to the delivery of services.We analyze, with greater depth, one of the lean techniques: the Value Stream Mapping. of this instrument are, first of all, highlighted the advantages and principles on which it rests, then is a description of the flow maps, the process boxes used and possible reconfiguration of a production system following the implementation of this instrument.

2   Case Study: A Healthcare Service in the South of Italy

With reference to the concepts learned in previous chapters, we have applied to real case, so that we can apply the theories to supply chain pre-existing views. The choice made for this analysis is that of a public service. We have focused our attention on a specific sector: the vaccination centers in the Public Health District.

2.1   Structural and Instrumental Organization of the Vaccination Centers

The immunization activities has provided a single room, inside the district and allocated to the first floor of the building. The room arranged is small/medium and in it all activities take place. This includes equipment that is equipped with two refrigerators of different capacities, a computer workstation for updating the software containing the vaccination cards of patients, and equipment necessary for administration. The staff consists of: a medical officer to vaccination and two nurses who deal daily updating of stock and vaccination cards in a paper and computerized formats, and a head nurse that interfaces with the District Pharmacy.

3   Application of Value Stream Mapping to Vaccination Services

To represent process we define different entities involved in vaccination services, grouped into one or more macro work.

We remember, in particular, that the service is the result of the integration of vaccination five macro-activities:

- Defining objectives and choice pharmaceutical company;
- Orders and Deliveries (Supply);
- Implementation of the vaccination protocol;
- Monitoring adverse events;
- Recovery not vaccinated.

The work starts by map processes and the result is a Current State Map that includes the physical and information flows. Then we will proceed first to analyzing the value operations of the map and at looking for any critical activities, and, then, we will create Future State Map, that defines every suggestions implemented to improve the "critical tasks".

3.1   Current State

First step to take is to realize the Value Stream Map of the current state. In this step will be highlighted those that are physical flows, represented by the delivery of the product, and information flows, easily identifiable with the supply orders. We start to identify different companies involved: **The vaccination centers**, is located within the local district; take place in the vaccination centers, almost all vaccine-related activities, from providing the administration; **the Pharmacy District**, a few miles from City, and offers its service, in addition to the vaccination centers even at different districts in the surrounding areas; **the CD** (Regional Healthcare) is a company with the functions of purchase and supply goods and equipment of health care; **suppliers** (Pharmaceutical).

- **Analysis of the Current State Map**

In this scheme can represent all those that are the crucial information about the processes, such as:

- All phases of physical transfer of vaccines
- Information relating to each phase
- Breakpoints flow
- Areas where the material is stored
- The average time relating to permanence in the various areas
- The information flow from the consumer to the supplier

The mapping, as anticipated, is the combination of the two maps, respectively, for the information flow and then to the physical. The product of this union is therefore an Extended Value Stream Map of the current state that looks like graphically: to simplify the flow and ensure a greater simplicity in the management of materials have been established some guidelines for a possible intervention:

- reduce the overall throughput time of the product within the value stream
- reduce the use of labor
- reduce the size of the stocks in warehouses interoperational.

We extended the mapping to integrate the information flow and physical with Value Stream Map of Pharmacy District and part of the vaccination centers. The objective of this paper is to highlight, more accurately, what is the link between the various activities and their performance. The physical flow of Pharmacy District is:

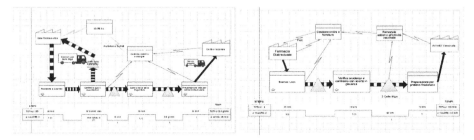

Fig. 1. District Pharmacy and Vaccination Center - Current State Maps

- **Finding the critical issues and proposals for improvement**

The identification of critical issues are those that are done following the guidelines of the Lean philosophy, we have already seen previously, to streamline the flow and ensure better management of the same. We think that if we add the supermarket that use pull logic and a milk run among different phase we can verify a more efficient operation.

3.2   Future State Map

After identifying the problems, the mapping activity requires that these process are highlighted on the first map of the current state, so that visually explain their position within the stream.

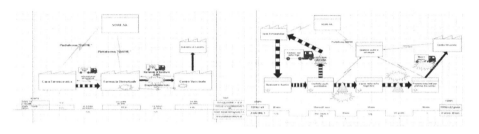

Fig. 2. The criticalities in the District Pharmacy

After identifying the critical areas where they are planted, we now proceed to the development of the same on the basis of what are the solutions previously identified. The different mappings that follow, therefore, will be the Future State Maps of the concerned activities.

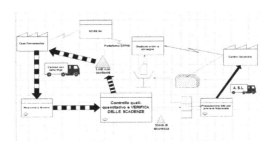

Fig. 3. District Pharmacy -Future State Map

From the analysis of the mapping can be seen the introduction of "supermarket" downstream processes followed by the activity of "preparation for lots withdrawal split" and from this activity on, as we shall see later, we assumed that the flow is continuous up to the storage in the vaccination centers. The latter interacts with the Pharmacy District through the use of kanban signals. Another suggestions for improvement represented. is to merge into a single activity the quail-quantitative control of batches and date monitoring. Delete the task of monitoring from the competence of the vaccination centers provides a streamlined flow, so that is continuous up to the storage in cold city rooms, and a reduction of lead time needed for such activities. The Future State Map of vaccination centers , as anticipated, shows the removal of the control of timing and implementation of a continuous flow until the storage of lots in two cold rooms of the Centre. Final note is that of the transport system from the Pharmacy District to vaccination centers. In the Future State Maps is optimistically assumed the use of a service financed by the public healthcare center to fulfill the delivery activities.

- **Simulation of corrective actions**

Downstream of the Future State Map we added a simulation phase in order to see if we had some improvement of some critical parameters of performance. The first step to take when you decide to start a simulation phase is to develop a model that describes as closely as possible the real situation emphasizing any critical present. Please note that the elaboration of the model have highlighted the activities that take place in both the District Pharmacy and vaccination centers in connecting with each other in that any critical encountered implications across the supply chain and, therefore, any improvement interventions can influence activities before the critical found.

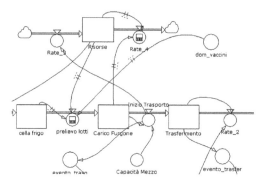

Fig. 4. vaccines delivery -current model (enlarged)

4   Conclusion

The Future State Map provides a visual representation of what are the improvements previously assumed. An analysis of the mapping can be seen the introduction of the "Supermarket" downstream processes followed by the activity of "preparation for sampling fractional lots." From this activity on, as we shall see later, we hypothesized that the flow is continuous up to the storage in the vaccination centers. The latter interacts with the Pharmacy District through the use of kanban signals, typical of Visual Management for the collection of the lots from the cells of the new system pull. These modification that we can implemented are tested by using simulation approach. In other words we have hypothesized new configuration on the process network, based on standard Lean Principles knowledge, and then we have verified, through modeling and simulation in the real functioning, the proposed evolutions.

References

1. Melton, T.: The Benefits of Lean Manufacturing: What Lean Thinking has to offer to Process Industries. Chemical Engineering Research and Design 83(A6), 662–673 (2005)
2. Holweg, M.: The Genealogy of Lean Production. Journal of Operations Management 25, 420–437 (2007)
3. Davis, M.M., Heineke, J.: Operations Management: integrating Manufacturing and Services, 5th edn., pp. 349–367. Mc-Graw Hill (2004)
4. Spear, S., Bowen, H.K.: Decoding the DNA ofthe Toyota Production System. Harvard Business Review 77(5) (September/October 1999)
5. Ohno, T., Bodek, N.: Toyota Production System: Beyond Large-Scale Production. Productivity Press, Portland (1988)
6. Saaksvuori, A., Immonen, A.: Product Lifecycle Management, 3rd edn. Springer (2008)
7. Arbòs, L.C.: Design of a Rapid Response and High Efficiency Service by Lean Production Principles: Methodology and Evaluation of Variability of Performance. International Journal of Production Economics 80, 169–183 (2002)

8. Ben-Tovim, D.I., Bassham, J.E., Bolch, D.M.M.A., Dougherty, M., Szwarcbord, M.: Lean Thinking Across a Hospital: Redesigning Care at the Flinders Medicai Centre. Australian Health Review 31(1) (2007)
9. Dickinson, E.W., Singh, S., Cheung, D.S., Wyatt, C.C., Nugent, A.S.: Application of Lean Manufacturing Techniques in the Emergency Department. The Journal of Emergency Medicine 37(2), 177–182 (2009)
10. Reitel, D.R., Rudkin, S.E., Malvehy, M.A., Killeen, P., Pines, M.: Improving Service Quality by Understanding Emergency Department Flow: A White Paper and Position Statement Prepared for the American Academy of Emergency Medicine. The Journal of Emergency Medicine 38(1), 70–79 (2010)
11. Ben-Tovim, D.I., Bassham, J.E., Bolch, D.M.M.A., Dougherty, M., Szwarcbord, M.: Lean Thinking Across a Hospital: Redesigning Care at the Flinders Medicai Centre. Australian Health Review 31(1) (2007)
12. Jones, D., Mitchell, A.: Lean Thinking for the NHS. NHS Confederation (2006)
13. Fillingham, D.: Can Lean Save Lives? The Journal of Leadership in Health Services 20(4), 231–241 (2007)
14. DeMaria, A.N.: The department of cardiac/ vascular medicine and surgery. J. Am. Coll. Cardiol. 46, 728–729 (2005)
15. Institute of Medicine: Crossing the Quality Chasm: a new health system for the 21st century. National Academy Press, Washington, DC (2001)
16. Laursen, M.L., Gertsen, F., Johansen, J.: Applying lean thinking in hospitals; exploring implementation difficulties. Aalborg University. Center for Industrial Production, Aalborg (2003)
17. Young, T.P., McClean, S.I.: A critical look at lean thinking in healthcare. Qual. Saf. Health Care 17, 382–386 (2008)
18. Elkhuizen, S.G., Limburg, M., Bakker, P.J., Klazinga, N.S.: Evidence-based re-engineering: re-engineering the evidence - a systematic review of the literature on business process redesign (BPR) in hospital care. Intern. J. Health Care Qual. Assur. Inc. Leadersh Health Serv. 19, 477–499 (2006)
19. Braithwaite, J., Westbrook, M.T., Hindle, T., Iedema, R.A., Black, D.A.: Does restructuring hospitals result in greater efficiency? – an empirical test using diachronic data. Health Serv. Manage Res. 19, 1–12 (2006)
20. Joosten, T., Bongers, I., Janssen, R.: Application of lean thinking to health care: issues and observations. Int. J. Qual. Health Care 21, 341–347 (2009)
21. Harrison, M., Henriksen, K., Hughes, R.G.: Improving the health care work environment: a sociotechnical systems approach. Jt Comm. J. Qual. Patient Saf. 33(suppl. 11), 3–6 (2007)
22. Aiken, L.H., Clarke, S.P., Sloane, D.M., et al.: Nurses' reports on hospital care in five countries. Health Aff. 20, 43–53 (2001)
23. Aiken, L.H., Clarke, S.P., Sloane, D.M.: Hospital restructuring: does it adversely affect care and outcomes? J. Nurs Adm. 30, 457–465 (2000)
24. Berwick, D.M.: Developing and testing changes in delivery of care. Ann. Intern. Med. 128, 651–656 (1998)
25. Winch, S., Henderson, A.J.: Making cars and making health care: a critical review. Med. J. Aust. 191, 28–29 (2009)

Decision Support Systems

Rationality for Adaptive Collective Decision Making Based on Emotion-Related Valuing and Contagion

Tibor Bosse[1], Jan Treur[1], and Muhammad Umair[1,2]

[1] VU University Amsterdam, Agent Systems Research Group
De Boelelaan 1081, 1081 HV Amsterdam, The Netherlands
[2] COMSATS Institute of Information Technology, Lahore, Pakistan
{t.bosse,j.treur}@vu.nl, mumair@ciitlahore.edu.pk

Abstract. This paper addresses a collective decision model based on interacting adaptive agents that learn from their experiences by a Hebbian learning mechanism. The decision making process makes use of emotion-related valuing of decision options on the one hand based on predictive loops through feeling states, and on the other hand based on contagion. The resulting collective decision making process is analysed from the perspective of learning speed and rationality. It is shown how the collectiveness amplifies both learning speed and rationality of the decisions.

Keywords: collective decision making, adaptive agent model, emotion-related valuing, Hebbian learning, rationality.

1   Introduction

Usually adaptive agents base their decisions on earlier experiences, using valuations and feelings for the decision options. In this way decisions are tuned to the environment and become more rational over time with respect to the world characteristics; e.g., [24]. Collective decision making in groups of agents combines these individual decision making processes with mutual contagion processes. In recent years some of the mechanisms underlying such social processes have been described in the area of Social Neuroscience (e.g., [9]). Two main concepts in these processes are mirror neurons and internal simulation. Mirror neurons are neurons that are not only active to prepare for a certain action or body change, but also when somebody else who is performing or tending to perform this action or body change is observed (e.g., [15, 21]). Internal simulation is mental processing that copies processes that (may) take place externally; e.g., [2, 6, 7, 11, 13, 17]). More specifically, before a decision option is chosen, by an internal simulation the expected effects of the option are predicted. For these predicted effects valuations are made, in relation to the affective state associated to this effect (e.g., [1, 6, 8, 16, 18]). To achieve a collective decision, in addition the feelings for a considered option by different agents affect each other: a form of contagion by which a decision option can get a shared positive emotion-related valuation, which can be the basis of a common decision. To analyse the effect of the collectiveness on the decision process and its rationality is the focus of the

W. Ding et al. (Eds.): Modern Advances in Intelligent Systems and Tools, SCI 431, pp. 103–112.

current paper. The decision making within the agent model involving valuing of the decision options by predictive valuation through feeling states, is adopted from [3, 24]; this decision model is based on cognitive and neurological literature such as [6, 8, 14, 15, 16, 17, 18, 19, 20]). The adaptation model used is based on Hebbian learning (cf. [10, 12]).

In this paper, first in Section 2 the multi-agent model is introduced. Section 3 presents the adaptation model based on Hebbian learning. Finally, Section 4 is a discussion. A number of appendices are available on Internet[1]. Simulation results are presented in Appendix A. In Appendix B two measures for rationality used are presented. Appendix C presents a mathematical analysis addressing equilibria, and the comparison of the learning speed for the cases with and without contagion. Appendix D describes tests of the learning speeds for some of the generated simulation traces and compares this with predictions made on the basis of the mathematical analysis. The adaptive joint decision making models are evaluated based on the rationality measures in Appendix E.

2   The Basic Model for the Agents and Their Contagion

This section describes the basic agent model used. Part of this model has been adopted from [24], which addresses especially the single agent case. The contagion mechanism has been added to this, and was adopted from [14].

2.1   Decision Making Based on Emotion-Related Valuing

In this part of the model emotional responses triggered by the environment play a role; see Fig. 1 for an overview. More specifically, it is assumed that responses in relation to a sensory representation state of a world stimulus w roughly proceed according to the following causal chain for a *body loop* (based on elements from [4, 7, 8]):

sensory representation $\rightarrow$ preparation for bodily response $\rightarrow$ body modification $\rightarrow$
sensing body state $\rightarrow$ sensory representation of body state $\rightarrow$ induced feeling

In addition, an *as-if body loop* uses a direct causal relation

preparation for bodily response $\rightarrow$ sensory representation of body state

as a shortcut in the causal chain; cf. [7]. This can be considered a prediction of the action effect by internal simulation (e.g., [13]). The resulting induced feeling provides an emotion-related valuation of this prediction (cf. [1, 6, 8, 16, 18, 20, 22, 23]). If the level of the feeling (which is assumed positive here) is high, a positive valuation is obtained. This valuation has a reinforcing effect on the preparation state. Therefore the body loop (or as-if body loop) is extended to a recursive (as-if) body loop by assuming that the preparation of the bodily response is also affected by the level of the induced feeling:

induced feeling $\rightarrow$ preparation for the bodily response

---

[1] URL http://www.few.vu.nl/~wai/IEA12collectiverationality.pdf [25]

Such recursion is also suggested in [8], pp. 91-92.Through this recursive loop a high valuation will strengthen activation of the preparation, which can lead to a high activation level as a basis for a decision for the option.

2.2   The Role of Contagion in a Collective Decision

Within the collective decision making model an additional mechanism for contagion has been incorporated, based on mirroring of the preparation states (adopted from [14]). An important element is the contagion strength γ_{BA} from person B to person A. This indicates the strength by which a preparation state S (for an option b_i) of A is affected by the corresponding preparation state S of B. It depends on characteristics of the two persons: how expressive B is, how open A is, and how strong the connection from B to A is. In the model it is defined by

$$\gamma_{BA} = \varepsilon_B \alpha_{BA} \delta_A$$

Here, ε_B is the *expressiveness* of B, δ_A the *openness* of A, and α_{BA} the *channel strength* from B to A. Note the labels in Fig. 1 for these concepts. The level q_{SA} of preparation state S in agent A (with values in the interval $[0, 1]$) over time is determined as follows. The overall contagion strength γ_A from the rest of the group towards agent A is $\gamma_A = \Sigma_{B \neq A} \gamma_{BA}$. The aggregated impact $q_{SA}*$ of all these agents upon state S of agent A is the following weighted average:

$$q_{SA}*(t) = \Sigma_{B \neq A} \gamma_{BA} q_{SB}(t) / \gamma_A$$

This is an additional external impact on the preparation state S of A, which has to be combined with the impact from the internal emotion-related valuing process. Note that for the case that there is only one other agent, this expression for $q_{SA}*(t)$ can be simplified to $q_{SB}(t)$.

Informally described theories in scientific disciplines, for example, in biological or neurological contexts, often are formulated in terms of causal relationships or in terms of dynamical systems. To adequately formalise such a theory the hybrid dynamic modelling language LEADSTO has been developed that subsumes qualitative and quantitative causal relationships, and dynamical systems; cf. [4]. This language has been proven successful in a number of contexts, varying from biochemical processes that make up the dynamics of cell behaviour to neurological and cognitive processes e.g. [4, 5]. Within LEADSTO the *dynamic property* or temporal relation a $\rightarrow_D$ b denotes that when a state property a occurs, then after a certain time delay (which for each relation instance can be specified as any positive real number D), state property b will occur. Below, this D will be taken as the time step Δt. In LEADSTO both logical and numerical calculations can be specified in an integrated manner, and a dedicated software environment is available to support specification and simulation. A formal specification of the model in LEADSTO can be found in current section.

An overview of the multi-agent model is depicted in Fig. 1. This picture also shows representations from the detailed specifications explained below. However, note that the precise numerical relations are not expressed in this picture, but in the detailed specifications below, through local properties LP0 to LP11.

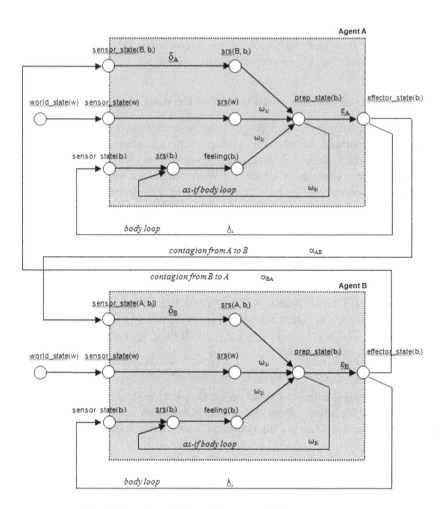

Fig. 1. Overview of the multi-agent model: two agents

The effector state for b_i combined with the (stochastic) effectiveness of executing b_i in the world (indicated by the *world characteristic* or *effectiveness rate* λ_i between 0 and 1) activates the sensor state for b_i via the body loop as described above. By a recursive as-if body loop each of the preparations for b_i generates a level of feeling for b_i which is considered a valuation of the prediction of the action effect by internal simulation. This in turn affects the level of the related action preparation for b_i. Dynamic interaction within these loops results in equilibrium for the strength of the preparation and of the feeling, and depending on these values, the action is actually activated with a certain intensity. The specific strengths of the connections from the sensory representation to the preparations, and within the recursive as-if body loops can be innate, or are acquired during lifetime. The computational model is based on such neurological notions as valuing in relation to feeling, body loop and as-if body loop, as briefly discussed above. In this paper the considered adaptation mechanisms for the model is based on Hebbian learning (Section 3). The detailed specification of the basic model is presented below, starting with how the world state is sensed.

LP0 Sensing a world state
If   world state property w occurs of level V
then   the sensor state for w will have level V.
  world_state(W, V) $\twoheadrightarrow$ sensor_state(W, V)

From the sensor state a sensory representation of the world state is generated by dynamic property LP1.

LP1 Generating a sensory representation for a sensed world state
If   the sensor state for world state property w has level V,
then   the sensory representation for w will have level V.
  sensor_state(W, V) $\twoheadrightarrow$ srs(W, V)

The combination function h to combine two inputs which activates a subsequent state is used along with the threshold function th to keep the resultant value in the interval [0, 1] as follows:

$$h(\sigma, \tau, V_1, V_2, \omega_1, \omega_2) = th\,(\sigma, \tau, \omega_1 V_1 + \omega_2 V_2)$$

where V_1 and V_2 are the current activation level of the states and ω_1 and ω_2 are the connection strength of the link between the states; here

$$th(\sigma, \tau, V) = \left(\frac{1}{1+e^{-\sigma(V-\tau)}} - \frac{1}{1+e^{\sigma\tau}}\right) * (1 + e^{-\sigma\tau})$$

where σ is the steepness and τ is the threshold of the given function. Alternatively for higher values of $\sigma\tau$, the threshold function $th(\sigma, \tau, V) = \frac{1}{1+e^{-\sigma(V-\tau)}}$ can be used. Dynamic property LP2 describes the generation of the preparation state from the sensory representation of the world state and the feeling thereby taking into account mutual inhibition. Here the combination function is defined as:

$$g(\sigma, \tau, V_1, V_2, V_3, V_4, \omega_1, \omega_2, \theta_{ji}, \theta_{ki}) = th\,(\sigma, \tau, \omega_1 V_1 + \omega_2 V_2 + \theta_{ji} V_3 + \theta_{ki} V_4)$$

where θ_{mi} is the strength of the inhibition link from preparation state for b_m to preparation state for b_i (which have negative values).

LP2 From sensory representation and feeling to preparation of a body state
If   a sensory representation for w with level V occurs
and   the feeling associated with body state b_i has level V_i
and   the preparation state for each b_m has level U_m
and ω_{1i} is the connection strength from sensory representation for w to preparation for b_i
and ω_{2i} is the strength of the connection from feeling of b_i to preparation for b_i
and θ_{mi} is for each m the strength of the inhibiting connection from preparation state for b_m to preparation state for b_i

and σ_i is the steepness value for preparation of b_i
and τ_i is the threshold value for preparation of b_i
and γ_1 is the person's flexibility for bodily responses
and $j{\neq}i,\ k{\neq}i,\ j{\neq}k$
then   after Δt the preparation state for body state b_i will have
    level $U_i + \gamma_1\ (g(\sigma_i,\ \tau_i, V,\ V_i,\ U_j,\ U_k,\ \omega_{1i},\ \omega_{2i},\ \theta_{ji},\ \theta_{ki})\text{-}\ U_i)\ \Delta t$

srs(w, V) & feeling(b_i, V_i) & $\wedge_m$ preparation_state(b_m, U_m) &
has_connection_strength(srs(w), preparation(bi), ω_{1i}) &
has_connection_strength(feeling(b_i), preparation(b_i), ω_{2i}) &
$\wedge_{m{\neq}i}$ has_connection_strength(preparation_state(b_m), preparation(b_i), θ_{mi}) &
has_steepness(prep_state(b_i), σ_i) & has_threshold(prep_state(b_i), τ_i) & j$\neq$i, k$\neq$i, j$\neq$k
$\rightarrow$ preparation(b_i, $U_i + \gamma_1$ (g(σ_i, τ_i,V, V_i , U_j, U_k, ω_{1i}, ω_{2i} , θ_{ji}, θ_{ki})- U_i) Δt)

Dynamic property LP3 describes the generation of sensory representation of a body state from the respective preparation state and sensory state.

LP3 From preparation and sensor state to sensory representation of a body state
If   preparation state for b_i has level X_i and the sensor state for b_i has level V_i
and the sensory representation for b_i has level U_i
and ω_{3i} is the strength of the connection from preparation state for b_i to sensory representation for b_i
and σ_i is the steepness value for sensory representation of b_i
and τ_i is the threshold value for sensory representation of b_i
and γ_2 is the person's flexibility for bodily responses
then   after Δt the sensory representation for body state b_i will have
    level $U_i + \gamma_2\ (h(\sigma_i,\ \tau_i, X_i,\ V_i,\ \omega_{3i},\ 1)\text{-}\ U_i)\ \Delta t.$

prep_state(b_i, X_i) & sensor_state(b_i, V_i) & srs (b_i, U_i) &
has_connection_strength(prep_state(bi), srs(bi), , ω_{3i}) &
has_steepness(srs(b_i), σ_i) & has_threshold(srs(b_i), τ_i)
$\rightarrow$ srs(b_i, $U_i + \gamma_2$ (h(σ_i, τ_i,X_i, V_i, ω_{3i}, 1)- U_i) Δt)

Dynamic property LP4 describes how the feeling is generated from the sensory representation of the body state.

LP4 From sensory representation of a body state to feeling
If   the sensory representation for body state b_i has level V,
then b_i will be felt with level V.
    srs(bi, V) $\twoheadrightarrow$ feeling(bi, V)

LP5 describes how an effector state is generated from respective preparation state, thereby taking into account the expressiveness ε_A.

LP5 From preparation to effector state
If   the preparation state for b_i has level V,   and ε_A is the expressiveness
then   the effector state for body state b_i will have level$\varepsilon_A V$.
    prep_state(bi, V) $\rightarrow$ effector_state(bi, ε_AV)

Dynamic property LP6 describes how a sensor state is generated from an effector state.

LP6 From effector state to sensor state of a body state

If   the effector state for b_i has level V_i,

and λ_i is the world characteristics for the option b_i

then the sensor state for body state b_i will have level $\lambda_i V_i$

effector_state(bi,V) & has_contribution(effecor_state(bi), sensor_state(bi),λ$_i$)

→ sensor_state(bi, λ$_i$V$_i$)

The following two properties model the contagion mechanism, first in LP7 from effector state of one agent to a sensor state of another agent (taking into account the channel strength), and next the further internal processing in the form of a sensory representation (taking into account the openness δ_A).

LP7 From effector state to sensor state of another agent (contagion)

If   the effector state for b_i has level V_i,   and α_{BA} is the channel strength

then the sensor state for body state b_i will have level $\alpha_{BA} V_i$

effecor_state(bi,V) & has_contribution(effecor_state(bi), sensor_state(bi),α_{BA})

→ sensor_state(B, bi, $\alpha_{BA}V_i$)

LP8 Generating a sensory representation for a sensed state of another agent

If   the sensor state for state property b_i of agent B has level V,

then the sensory representation for state property b_i of agent B will have level $\delta_A V$.

sensor_state(B, bi, V) → srs(B, bi, δ$_A$V)

For the case studies addressed three options are assumed available in the world for the agent. The objective is to see how rationally agents make collective decisions using the given model (under static as well as stochastic world characteristics).

3   An Adaptation Model by Hebbian Learning

From a Hebbian perspective [12], strengthening of a connection over time may take place when both nodes are often active simultaneously ('neurons that fire together wire together'). The principle goes back to Hebb [12], but has recently gained enhanced interest (e.g., [10]). In the adaptive computational model three connections that play a role in decision making can be adapted based on a Hebbian learning principle. More specifically, for such a connection from node i to node j its strength ω_{ij} is adapted using the following *Hebbian learning rule*, taking into account a maximal connection strength 1, a *learning rate* η, and an *extinction rate* ζ (usually taken small):

$$\frac{d\omega_{ij}(t)}{dt} = \eta a_i(t)a_j(t)(1 - \omega_{ij}(t)) - \zeta\omega_{ij}(t) = \eta a_i(t)a_j(t) - (\eta a_i(t)a_j(t) + \zeta)\,\omega_{ij}(t)$$

Here $a_i(t)$ and $a_j(t)$ are the activation levels of node i and j at time t and $\omega_{ij}(t)$ is the strength of the connection from node i to node j at time t. A similar Hebbian learning rule can be found in [10], p. 406. By the factor $1 - \omega_{ij}(t)$ the learning rule keeps the level of $\omega_{ij}(t)$ bounded by 1 (which could be replaced by any other positive number);

Hebbian learning without such a bound usually provides instability. When the extinction rate is relatively low, the upward changes during learning are proportional to both $a_i(t)$ and $a_j(t)$; maximal learning takes place when both are *1*. Whenever one of $a_i(t)$ and $a_j(t)$ is *0*(or close to*0*) extinction takes over, and ω_{ij} slowly decreases. This learning principle has been applied for example to the connection from sensory representation of *w* to the preparation state for b_i in Fig. 1, according to the following instantiation of the general learning rule above:

$$\frac{d\omega_{1iA}(t)}{dt} = \eta srs(w)(t)prep\_state(bi)(t)(1 -\omega_{1iA}(t)) - \zeta\omega_{1iA}(t)$$

In this section the basic agent model described in Section 2 is extended with a Hebbian learning approach. The connection strength of the different links are updated according to the local properties LP9 to LP11.

**LP9 Hebbian learning for connection from sensory representation of stimulus
       to preparation of b_i**

If   the connection from sensory representation of w to preparation of b_i has strength ω_{1i}
  and the sensory representation for w has level *V* and the preparation of *bi* has level V_i
  and the learning rate from sensory representation of w to preparation of b_i is η
  and the extinction rate from sensory representation of w to preparation of b_i is ζ
       then   after Δt the connection from sensory representation of w to preparation of *bi*   will
       have strength $\omega_{1i} + (\eta VV_i(1 - \omega_{1i}) - \zeta\omega_{1i}) \Delta t$.
  has_connection_strength(srs(w), preparation(b_i), ω_{1i}) &srs(w, V) & preparation(b_i, V_i) &
  has_learning_rate(srs(w), preparation(b_i), η) &
  has_extinction_rate(srs(w), preparation(b_i), ζ)
  $\rightarrow$ has_connection_strength(w, b_i, $\omega_{1i} + (\eta VV_i(1 - \omega_{1i}) - \zeta\omega_{1i}) \Delta t$)

Similarly, LP10-LP11 (not shown) specify Hebbian learning for the connections from feeling to preparation, and from preparation to sensory representation of b_i.

Simulation experiments were performed using numerical software. Learning of the connections was done either one at a time, or for multiple connections simultaneously. To simulate a stochastic world, probability distribution functions (PDF) were defined for the world characteristic parameters λ_i according to a normal distribution. For more details, see the appendices in [25].

4   Discussion

The presented adaptive collective decision model is based on interacting adaptive agents that learn from their experiences by a Hebbian learning mechanism (cf. [10, 12]). Within each agent, the decision making process makes use of (1) emotion-related valuing of decision options by internal simulation, and (2) social contagion processes. The internal simulations are based on predictive loops through feeling states, inspired by literature on the neurological basis of valuing; e.g., [1, 6, 8, 16, 18, 20, 22, 23].

The resulting collective decision making process was analysed on learning speed, and on rationality of the collective decisions after a major world change. This was done by a mathematical analysis providing approximated learning speeds, and based on analysis of simulation results. By both types of analysis it was shown how the contagion amplifies the learning speed, and therefore strengthens the rationality of the decision making, in particular after changes of the world characteristics.

References

1. Bechara, A., Damasio, H., Damasio, A.R.: Role of the Amygdala in Decision-Making. Ann. N.Y. Acad. Sci. 985, 356–369 (2003)
2. Becker, W., Fuchs, A.F.: Prediction in the Oculomotor System: Smooth Pursuit During Transient Disappearance of a Visual Target. Experimental Brain Res. 57, 562–575 (1985)
3. Bosse, T., Hoogendoorn, M., Memon, Z.A., Treur, J., Umair, M.: An Adaptive Model for Dynamics of Desiring and Feeling Based on Hebbian Learning. In: Yao, Y., Sun, R., Poggio, T., Liu, J., Zhong, N., Huang, J. (eds.) BI 2010. LNCS (LNAI), vol. 6334, pp. 14–28. Springer, Heidelberg (2010)
4. Bosse, T., Jonker, C.M., Treur, J.: Formalisation of Damasio's Theory of Emotion, Feeling and Core Consciousness. Consciousness and Cognition Journal 17, 94–113 (2008)
5. Bosse, T., Jonker, C.M., Meij, L., van der Treur, J.: A Language and Environment for Analysis of Dynamics by Simulation. Intern. J. of AI Tools 16, 435–464 (2007)
6. Damasio, A.: Descartes' Error: Emotion, Reason and the Human Brain. Papermac, London (1994)
7. Damasio, A.: The Feeling of What Happens. Body and Emotion in the Making of Consciousness. Harcourt Brace, New York (1999)
8. Damasio, A.: Looking for Spinoza. Vintage books, London (2004)
9. Decety, J., Cacioppo, J.T. (eds.): Handbook of Social Neuroscience. Oxford University Press (2010)
10. Gerstner, W., Kistler, W.M.: Mathematical formulations of Hebbian learning. Biol. Cybern. 87, 404–415 (2002)
11. Goldman, A.I.: Simulating Minds: The Philosophy, Psychology, and Neuroscience of Mindreading. Oxford Univ. Press, New York (2006)
12. Hebb, D.O.: The Organization of Behaviour. John Wiley & Sons, NewYork (1949)
13. Hesslow, G.: Conscious thought as simulation of behaviour and perception. Trends Cogn. Sci. 6, 242–247 (2002)
14. Hoogendoorn, M., Treur, J., van der Wal, C.N., van Wissen, A.: Modelling the Interplay of Emotions, Beliefs and Intentions within Collective Decision Making Based on Insights from Social Neuroscience. In: Wong, K.W., Mendis, B.S.U., Bouzerdoum, A. (eds.) ICONIP 2010, Part I. LNCS (LNAI), vol. 6443, pp. 196–206. Springer, Heidelberg (2010)
15. Iacoboni, M.: Mirroring People: the New Science of How We Connect with Others. Farrar, Straus & Giroux, New York (2008)
16. Montague, P.R., Berns, G.S.: Neural economics and the biological substrates of valuation. Neuron 36, 265–284 (2002)
17. Moore, J., Haggard, P.: Awareness of action: Inference and prediction. Consciousness and Cognition 17, 136–144 (2008)
18. Morrison, S.E., Salzman, C.D.: Revaluing the amygdala. Current Opinion in Neurobiology 20, 221–230 (2010)

19. Murray, E.A.: The amygdala, reward and emotion. Trends Cogn. Sci. 11, 489–497 (2007)
20. Rangel, A., Camerer, C., Montague, P.R.: A framework for studying the neurobiology of value-based decision making. Nat. Rev. Neurosci. 9, 545–556 (2008)
21. Rizzolatti, G., Sinigaglia, C.: Mirrors in the Brain: How Our Minds Share Actions and Emotions. Oxford University Press (2008)
22. Salzman, C.D., Fusi, S.: Emotion, Cognition, and Mental State Representation in Amygdala and Prefrontal Cortex. Annu. Rev. Neurosci. 33, 173–202 (2010)
23. Sugrue, L.P., Corrado, G.S., Newsome, W.T.: Choosing the greater of two goods: neural currencies for valuation and decision making. Nat. Rev. Neurosci. 6, 363–375 (2005)
24. Treur, J., Umair, M.: On Rationality of Decision Models Incorporating Emotion-Related Valuing and Hebbian Learning. In: Lu, B.-L., Zhang, L., Kwok, J. (eds.) ICONIP 2011, Part III. LNCS(LNAI), vol. 7064, pp. 217–229. Springer, Heidelberg (2011)
25. http://www.few.vu.nl/~wai/IEA12collectiverationality.pdf

A Model of Team Decision Making Using Situation Awareness

Mark Hoogendoorn[1], Pieter Huibers[2], Rianne van Lambalgen[1], and Jan Joris Roessingh[2]

[1] VU University Amsterdam, Agent Systems Research Group
De Boelelaan 1081, 1081 HV Amsterdam, the Netherlands
{m.hoogendoorn,r.m.van.lambalgen}@vu.nl
[2] National Aerospace Laboratory
Training, Simulation & Operator Performance
Anthony Fokkerweg 2 Amsterdam, the Netherlands
{jan.joris.roessingh,pieter.huibers}@nlr.nl

Abstract. In order for agents to successfully make decisions about task allocations within a team, two elements are essential: (1) a good judgement of the situation, and (2) once the situation is known have a good decision making process to derive and assign tasks that should be performed. Within research on agent systems, little work has been done on the combination of the two. In this paper, a human-based situation awareness model is combined with a decision making procedure (which incorporates task identification and task allocation).

1 Introduction

In order for agents to be able to perform well in decision making, one crucial aspect is that such agents have sufficient knowledge about the situation. Endsley (1997) states that "in most settings effective decision making depends on having a good understanding of the situation at hand". This "understanding of the situation at hand" is often referred to as situation awareness (see Endsley, 1997). Of course, once good situation awareness is established, the effectuation of a decision is not a trivial matter either, especially in settings whereby multiple agents play a role, for instance when agents cooperate in teams. Crucial elements hereby include the allocation of tasks to agents as well as the monitoring of the progress of their task execution (and thus constantly maintaining a good understanding of the situation).

In this paper, an integrated framework is presented for situation awareness in combination with decision making in teams, which is novel in the domain of agent systems. Hereby, the model has been created for domains where centralized decision making is adopted within the team (i.e. one agent is in charge). The framework is inspired upon the theory of Klein (1998) which indicates five important cognitive aspects in coordinated team decision making: (1) Control of attention; (2) Shared situation awareness; (3) Shared mental models; (4) Application of strategies and heuristics to make decisions, solve problems, and plan; (5) Theory of Mind (or meta

W. Ding et al. (Eds.): Modern Advances in Intelligent Systems and Tools, SCI 431, pp. 113–120.
springerlink.com   © Springer-Verlag Berlin Heidelberg 2012

cognition, i.e. thinking about (other team members' or opponents') thinking). The model has been evaluated in the domain of fighter pilots (more specifically in air-to-air combat situations). In this context, team based decision making is modelled as a process that aims at the combined use of aircraft and weapons in order to defeat or gain advantage over an adversary.

This paper is organized as follows. In Section 2 the model is presented. Section 3 presents the result of the application of the model upon the case study. Finally, Section 4 is a discussion.

2   Multi-Agent System

Section 2 has been used as a source of inspiration to develop a model for team decision making. The overall architecture consists of two types of agents: (1) a task allocation agent, and (2) a task execution agent. The former agent is composed of four components. First of all, a component is present which creates a judgment of the current situation (the *situation awareness* component). Thereafter, this information is forwarded to the component *tactic selection*, which takes this situation into account when selecting the most appropriate tactic. The tactic is then forwarded to the *task determination* component which derives the tasks to be executed given this tactic. Finally, the component *task allocation* determines which agents to allocate to these tasks. This task allocation is then forwarded to the appropriate agents. These in turn execute the task (i.e. the *task execition agent*), for which no specific restrictions are imposed on the itnernal structure of the agents.

Below, each of the components within the task allocation agent is explained in more detail.

Situation Awareness. The first element in the model of the task allocation agent involves the creation of a good situation awareness within the agent in charge as also highlighted in Section 2. In order to do so, the model presented in Hoogendoorn, van Lambalgen, and Treur (2011) has been adopted which is based upon the theory proposed by Endsley (1995). In the theory, situation awareness is said to include the perception of cues, the comprehension and integration of information and the projection of information for future events. For the sake of brevity, the full details of the model are not explained (see Endsley, 1995 for these details), but essentially, the model comprises of four parts:

1. Formation of simple beliefs about the current situation based upon observations and communications.
2. Integration of beliefs into more complex aggregated beliefs about the current situation.
3. Generation of future beliefs based upon the complex beliefs.
4. Determination of observations and communications to be performed given the current set of complex beliefs and future beliefs (communication is used to create a good shared situation awareness as expressed by Klein (1998)).

In order to perform these steps, knowledge is present within the model that expresses how the various elements are connected: what observations lead to simple beliefs, how simple beliefs influence each other, how complex beliefs are influenced by simple beliefs, and how complex beliefs result in future beliefs. These influence relations are expressed by means of weights of connections between these elements. An algorithm then derives the new activation values of the states based upon the incoming observations and the old values of the states in combination with the weights as described above. This algorithm has anytime behaviour. The activation levels of the complex beliefs act as judgment of the current situation and are output of this component.

Tactic Selection. The second component concerns the decision making process to decide upon a tactic to follow and essentially represents the fourth aspect identified by Klein as put forward in Section 1 (i.e. application of strategies and heuristics to make decisions, solve problems, and plan). For this purpose, the following decision rule is used inspired by a decision making model as proposed in (Hoogendoorn, Merk, and Treur, 2010). The model essentially encompasses the following steps:

1. Generate possible tactics from the currently active complex beliefs describing the situation.
2. Generate a feeling associated with each option (called a Somatic Evaluation Value, cf. (Damasio, 1994)) by means of the weights assigned to the goals the agents have and the feeling associated with each of the options with respect to that goal (by taking a weighed sum per option of the values with respect to the goals and the weights of the goals).
3. Generate a rational utility for each of the options.
4. Choose an option by taking the highest value of the weighed sum of the outcome of (2) and (3) per option. The weight of the rational and feeling part depends upon the rationality factor of the agent.

More details about the decision making process can be found in (Hoogendoorn, Merk, and Treur, 2010).

Task Determination. Once a tactic has been selected, a workflow needs to be derived that expresses what tasks accompany the specific tactic to be executed (again following aspect (4) identified by Klein). For this purpose, a workflow representation is used that expresses the ordering in which tasks should be performed within the tactic as well as the requirements associated with each of the tasks (i.e. requirements with respect to the allocation agent). Formally, each task is defined by means of the attributes listed in Table 1. The last element in the table is the output of the component, namely the schedule for the tasks. Note that the additional information is also passed on to enable the finding of a suitable task allocation.

Table 1. Task specification language

Predicate/Sort	Explanation
TASK	Representation of a task.
TIME	Representation of time.
REQUIREMENT	The task allocation requirements can be a complex domain dependent term.
task_preceded_by: TASK x TASK	A task is preceded by another task which should be completed before the task can start.
task_duration: TASK x TIME	The duration of a certain task.
task_allocation_requirement: TASK x REQUIREMENT	The task has a specific allocation requirement.
task_scheduled_from_to: TASK x TIME x TIME	A task is scheduled from a certain start time to a certain end time. Note that the difference should be equal to the task duration.

Task Allocation. Once the tasks have been expressed, an allocation should be found that fulfills the allocation requirements as expressed in the workflow. Of course, there might still be many possible configurations left. Therefore, as an additional criterion the utilization of resources is used (this could be both in terms of mental as well as physical resources). Table 2 expressed the information that is used by the component.

Table 2. Allocation information used

Predicate/Sort	Explanation
fulfills_requirement: AGENT x TASK x REQUIREMENT	The agent fulfills the requirements for the task (e.g. sufficient education).
current_resource_utilization: AGENT x RESOURCE x VALUE x TIME	The current utilization of a certain resource by the agent.
required_resource_utilization: AGENT x TASK x RESOURCE x VALUE	The amount of resources required for the agent to enable the execution of the task.
prospected_resource_utilization: AGENT x RESOURCE x VALUE x TIME	The prospected usage of a certain resource by the agent, given that certain additional tasks are assigned to the agent.
maximum_resource_utilization: AGENT x RESOURCE x VALUE	The limit of the agent's capabilities.
task_allocation: TASK x AGENT x TIME x TIME	An agent is allocated to the task during the specified interval.

Based upon this information, the set of possible allocations is determined. All combinations of agents to tasks are generated for which all agents fulfil the requirements of the tasks they have been allocated to. For the allocation itself, the following algorithm has been designed:

Algorithm 1. Determine allocation

```
For all possible allocations TA
      current_score(TA) = 0;
      For all agents that are part of task allocation TA
            For all resources R that the agent has
                  Calculate the prospected resource utilization given the current task allocations
                  and current resource utilization.
                        if the prospected resource utilization is above the maximum
                              Set current_score = high
                        else
                              Calculate the score on the evaluation function.
                              Add the score to current_score(TA).
                        end
            end
      end
end
Select the task allocation with the lowest value on the current score.
```

The algorithm essentially determines whether the task allocation does not result in any exceeding of resources (otherwise the score is set to a very high value in case it will only be selected if no other option besides the exceeding of resources is available) and calculates the score using a particular evaluation function. Various evaluation functions can be present. In this case, two evaluation functions are proposed: (1) the minimization of the overall resource utilization (then the score used above is equal to the prospected resource utilization), or (2) minimization of the percentage of utilization (the score equals the division of prospected resource utilization by the maximum value). Using such an evaluation function involves the reasoning about the other team members (i.e. a form of theory of mind) which is the fifth point made by Klein.

3   Results

In this section, first the case study is presented, followed by the implementation details. Finally, the results are presented.

Case Study. In order to evaluate the model proposed in Section 2, a case study has been conducted. This concerns a study in the domain of fighter pilots. In this particular case, there are four fighters active, i.e. a real, manned, formation (BLUE) and an agent controlled formation (RED), each formation consisting of a 'flight lead' and a 'wingman', and two opponents. It is assumed that both formations have aircraft that are capable of beyond-visual-range engagements. This, in turn, means that pilots may only 'see' each other via their on-board radar or via their Radar Warning Receiver (RWR). The latter device warns the pilot as soon as it receives radiation from the opponent's radar.   More specifically, the aircraft's RWR may signal the pilot that: (1) the opponent's radar is merely *searching* for targets, or (2) that the opponent's radar is actually *tracking* the aircraft, meaning that the position of the aircraft is tracked over a substantial period, which in turn may indicate that the opponent will shoot at the aircraft as soon as the opportunity arises.

When the two formations encounter each other, the leader and the wingman of the agent controlled RED formation should decide upon the tactical manoeuvres that they should execute, in this case manoeuvre A and manoeuvre B. These manoeuvres fall under the heading of air-to-air tactics. In order to select a tactic, dedicated domain information has been inserted as knowledge in the *situation awareness* component of the model (by means of elicitation from experts). This knowledge is e.g. whether the RED flight lead or wingman is searched or tracked by the BLUE formation and the spatial configuration of the BLUE formation (mutual separation, altitude difference). Based upon the judgment of the situation, the RED flight lead selects one out of three tactical options: (1) continue straight and level flight; (2) tactical manoeuvre A, and (3) tactical manoeuvre B. Each of these tactical manoeuvres consists of a number of elementary tasks: *fly_to_position(x,y) 180_degree_turn, 90_degree_right_turn, 90_degree_left_turn* and *continue_straight_and_level_flight*. These tactical manoeuvres were expressed in the form of a workflow as specified in Section 2. Furthermore, the requirements were composed in such a way that only one aircraft (the RED flight lead or the RED wingman) would be assigned to the task. These requirements involved the fact that the agents were being searched or tracked, and their relative positions.

Implementation. For this case study, the approach as presented in Section 3 was implemented in Matlab. In order to enable a simulation of the agents, a world model has also been implemented. The model essentially consists of the x and y position and the heading of the agents which change based upon the actions performed by the agents. For example, when an agent had to execute the task to go to position 1, the x and y coordinates of the agent were adjusted towards position 1. The world model also determines what observations the agent receives as input for the situation awareness model, this includes information about whether the agent is being searched by the opponent or tracked. This was made dependent upon the relative position of the fighter planes (distance and angle). A very simple radar model was implemented for this purpose:

$$p_{searched}(i, j) = \quad w_{distance}(1 - (distance(i, j) / search\_range)) +$$
$$w_{angle}(1 - (angle(i, j) / search\_angle))$$

For tracking the same equation has been used, except that tracking can only occur in case a successful search was encountered, and the range is set to a lower value. For more details on the case study, and the parameter settings used, (see Appendix A). Note that the scenario is not deterministic as the detection probabilities play an important role. For analysis, the behaviour of the agents within the multi-agent framework was compared to that of agents within a framework where tactics are randomly selected as well as a fixed continue flying tactic. More specifically, measurements were performed to see how many times the agents were searched and tracked as this is the performance indicator for such tactics. The scenario was executed 100 times. The results are presented below.

Results. For comparison of the three models (intelligent tactic, random tactic and fixed continue flying tactic), the total number of times that both agents were searched and the total number of times that both agents were tracked were calculated.

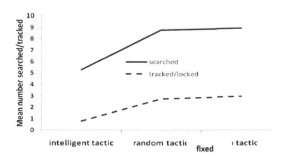

Fig. 1. Mean number of times that the flightlead and wingman are either searched or tracked/locked

In Figure 3 the results are graphically represented. A repeated measures analysis (Huynh-Feldt) was performed, which showed that the number of times the agents were searched was significantly less for the intelligent tactic selection ($M=5.29$ $SD=2.17$) as compared to the random tactic selection ($M=8.75$, $SD=1.87$) and fixed continue flying tactic selection ($M=8.94$, $SD=1.87$; $F(2, 99)=115.83$, $p<0.001$). Paired t-tests were conducted to compare the separate models. A difference was found between the intelligent tactic selection and the random tactic selection ($t(99)=-12.53$, $p<0.001$) as well as fixed tactic selection ($t(99)=-13.33$, $p<0.001$).

Also, analysis of the number of times that the agents were tracked showed that there was a significant difference between the intelligent tactic selection ($M=0.8$, $SD=0.98$), the random tactic selection ($M=2.73$, $SD=1.48$) and fixed tactic selection ($M=2.95$, $SD=1.47$; $F(2, 99)=74.58$, $p<0.001$). Again, the paired t-test that compared the separate selection strategies showed a significant difference between the intelligent tactic selection and the random tactic selection ($t(99)=-11.11$, $p<0.001$ and fixed tactic selection ($t(99)=-12.10$, $p<0.001$). These promising results show that a tactic selection based on the agent's situation awareness is effective.

4   Discussion

In this paper, a model for team decision making has been presented. The model has been inspired by various psychological theories (including those of Klein (1998) and Endsley (1995;1997) and consists of a number of components to come to judge the situation, derive a tactic and assign agents to the tasks that are part of the theory. As already indicated before, within the research in multi-agent systems, the combination of both models for situation awareness with more complex forms of team decision making has not been studied previously, whereas the importance thereof is highly emphasized by for instance Klein (1998). Within the domain of the case study, the model shows promising results.

Next steps for the research presented in this paper involve a more rigorous evaluation of the model itself. The main idea of this approach is to incorporate the model in a simulated opponent for fighter pilots that receive training in a flight simulator. The next step is then to evaluate the training value for these pilots, and investigate whether they consider the team decision making by there opponent's as intelligent.

References

Appendix A: http://www.cs.vu.nl/~mhoogen/sa/appendix_a.pdf

Damasio, A.: Descartes' Error: Emotion, Reason and the Human Brain. Papermac, London (1994)

Endsley, M.R.: Toward a theory of Situation Awareness in dynamic systems. Human Factors 37(1), 32–64 (1995)

Endsley, M.R.: The Role of Situation Awareness in Naturalistic Decision Making. In: Zsambok, C.E., Klein, G.A. (eds.) Naturalistic Decision Making, pp. 269–283. Lawrence Erlbaum Associates (1997)

Hoogendoorn, M., van Lambalgen, R., Treur, J.: Modeling Situation Awareness in Human-Like Agents using Mental Models. In: Walsh, T. (ed.) Proceedings of the Twenty-Second International Joint Conference on Artificial Intelligence, IJCAI 2011, pp. 1697–1704 (2011)

Hoogendoorn, M., Merk, R.J., Treur, J.: An Agent Model for Decision Making Based upon Experiences Applied in the Domain of Fighter Pilots. In: Huang, X.J., Ghorbani, A.A., Hacid, M.-S., Yamaguchi, T. (eds.) Proceedings of the 10th IEEE/WIC/ACM International Conference on Intelligent Agent Technology, IAT 2010, pp. 101–108. IEEE Computer Society Press (2010)

Klein, G.A.: Sources of Power: How People Make Decisions, pp. 1–30. MIT Press, Cambridge (1998)

Subject-Oriented Science Intelligent System on Physical Chemistry of Radical Reactions

Vladimir Tumanov[1] and Bulat Gaifullin[2]

[1] Institute of Problems of Chemical Physics RAS, Semenov ave 17,
142432 Chernogolovka, Russia
tve@icp.ac.ru
[2] Interface Ltd., Bardin str., 4/1 119334, Moscow, Russia
bulat@interface.ru

Abstract. In article the subject-oriented scientific intelligence system on physical chemistry of radical reactions is presented, its program and technological architectures are described, the structure of program components and their purpose is resulted.

1 Introduction

As a result of scientific experiments and computer modeling, large amounts of data are constantly accumulated and are organized in electronic information resources: databases and data warehouse, electronic information and computer systems, scientific data centers. Effective development of science and high technology requires intensive processing and analysis of the fundamental knowledge gained in various research organizations, which leads to the need in development of information technology of storage, retrieval and analysis of subject-oriented professional knowledge through the development of generic and specialized models of organization and presentation of scientific data and knowledge in electronic resources.

One of the approaches to solving the urgent problem for all professional communities of science, business and government, - storage, verification, retrieval, use and dissemination of professional knowledge - is the creation of scientific intelligence system [1-4]. [5] provides the definition of *subject-oriented science intelligence systems*, as *highly specialized science intelligence systems*, which in addition to the possibility of solving data mining tasks, are endowed with the ability to produce new professional knowledge.

The purpose of this paper is to attempt to determine the class of subject-oriented science intelligence systems and formulate the basic requirements for such systems, their overall software technology architecture.

2 Subject-Oriented Science Intelligence System in Physical Chemistry of Radical Reactions

Subject-oriented science intelligence system in physical chemistry of radical reactions is regarded as an intelligent system in the Internet, whose purpose is the collection,

W. Ding et al. (Eds.): Modern Advances in Intelligent Systems and Tools, SCI 431, pp. 121–126.
springerlink.com

storage, verification, retrieval, distribution and production of new subject-oriented knowledge of the physical chemistry of radical reactions.

The subject area of the system consists of the following objects and their main characteristics: organic molecule (chemical formula, name, description of the reaction center, CAS Number, type of the ruptured bond, bond dissociation energies, enthalpy of formation of a molecule); radical (chemical formula, name, description of the reaction center, CAS Number, the enthalpy of formation of the radical); radical reaction (reagents, reaction type, reactivity, phase, solvent, reaction conditions); bibliographic reference to the source of factual data;e-lecture for distance learning and tasks for self-control; a thesaurus of subject-oriented knowledge (definition of terms, context-sensitive hints).

To design and build the system, the approach based on portal technologies, technology of intellectual agents, technologies of applied artificial intellect, data mining technology, technology of databases and warehousing was used. The active component of the system is an intellectual agent that can be represented in the form of web - application located behind the external information portal. At that, these agents are focused on the processing of scientific data in the highly specialized section of subject domain. The overall software and technology architecture of the system is shown in Figure 1.

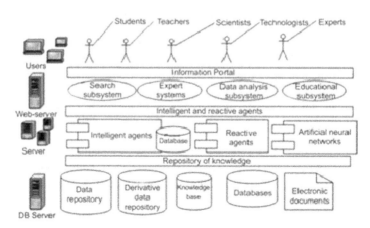

Fig. 1. The software and technology architecture of subject-oriented science intelligence system in physical chemistry of radical reactions

As shown on the Fig. 1, the system consists of several software layers. The first layer is implemented as an subject-oriented web - application that introduces the interface to the user and makes management decisions based on decision tables. This web - application gives access to the following software components of the system: information subsystem, analytical subsystem, distance learning subsystem, embedded subsystem of explanations and subsystem for producing new professional knowledge.

The second layer is composed of intellectual agents, reactive agents and trained artificial neural networks that implement the work of expert systems built into the portal and perform the functions of information retrieval. The agents are distributed in different nodes of the local network.

The third layer is the repository of knowledge, which consists of: experimental data warehouse on the rate constants of radical liquid-phase reactions, bond dissociation energies of molecules, the enthalpies of formation of molecules that are implemented as linked data marts (intensional data marts); derivative data warehouses on the rate constants of radical liquid-phase reactions, bond dissociation energies of molecules, the enthalpies of formation of molecules, in which users are allowed to enter the new values they have received as a result of working with the system (extensional data marts); the knowledge base, which consists of production rules, decision tables, procedures for performing calculations, cluster analysis algorithms, and the general facts used by the expert systems; electronic documents that represent, in particular, the materials of educational lectures on physical chemistry of radical reactions, as well as thesaurus of terms and explanation subsystem files.

The system is designed to solve the following tasks: search for the experimental data on the reactivity of reactants in bimolecular radical reactions in the liquid phase; search for the calculated data on the reactivity of reactants in the bimolecular radical reactions in the liquid and gas phases; search for the values of bond dissociation energies of organic molecules, as well as enthalpies of formation of radicals and molecules; search for bibliographic references in the database; evaluation of the reactivity of the reagents of radical bimolecular reactions in liquid and gas phases by the thermochemical and kinetic data; evaluation of bond dissociation energies of organic molecules by kinetic data.

3   Knowledge Warehouse

Knowledge warehouse, by definition, is an subject-oriented, integrated electronic collection supporting time-series data, which contains the data, knowledge, procedures for knowledge generation. It is used for data analysis and research, production of new knowledge and decision making support.

Knowledge warehouse of this system contains empirical and calculated facts, production rules and procedures for calculating, and in conjunction with expert systems constitutes a virtual subsystem for production of new professional knowledge (the constants of rate and energies of radical reactions activation, energies of molecular bonds dissociation).

Knowledge warehouse as a component of new professional knowledge production includes: Exploration Data Warehouse, which contains experimental data on the reactivity of radical reactions in the liquid phase; integrated expert system to manage the evaluation of reactivity of reagents radical reactions, representing a combination of intellectual and reactive agents); intellectual agent to evaluate the rate constants and activation energies of reactions in the liquid and gas phase; web service, through which the call of trained artificial neural networks is performed to predict the values of rate constants and activation energies of liquid-phase radical reactions of certain classes; derivative data warehouse containing the calculated data on the reactivity of radical reactions in liquid and gas phases; integrated expert system to evaluate the bond dissociation energy of molecular by kinetic data of radical reactions; data warehouse of bonds dissociation energies of organic molecules, which can be supplemented with new data as a result of work of expert system for evaluation of bonds dissociation energy of molecular by kinetic data of radical reactions; thesaurus of key terms and concepts in the subject domain; thesaurus of descriptions of algorithms and procedures for predicting physicochemical characteristics of molecules; knowledge base containing production rules and facts, which are used by integrated expert systems.

As a result of the work of the system users the knowledge warehouse is updated with new professional knowledge. As the mechanisms of production of new knowledge in this present system the expert systems integrated into the portal, trained artificial neural networks and intellectual agents are used.

As experimental operation of the system demonstrated, the use of the integrated expert systems is justified when it is possible to use expert knowledge or to obtain such knowledge through data mining, which is the time-consuming process.

The use of artificial neural systems, which in some cases better approximate the dependences of the data and provide a more accurate prediction of the reactivity, turned to be more promising. However, artificial neural networks are not always possible to be trained because of a lack of representative training set.

Information thesaurus of the system have been developed both manually and automatically using fuzzy clustering algorithms.

The information component of the system is the part of knowledge warehouse and consists of databases and data warehouses. Data warehouse of the system is implemented as linked data marts built using multi-dimensional modeling.

The information component of the system includes: data mart of the rate constants of radical liquid-phase reactions; data mart of bonds dissociations energies of organic molecules; data mart of the enthalpies of formation of organic molecules; a database of enthalpies of formation of radicals; a database of bibliographic references.

4   Production of New Knowledge

Production of new data within the subject domain of the science intelligence system being considered, is performed by means of expert systems. Expert Systems (ES), designed for operation in the Internet environment, are constructed as a set of intellectual software agents - autonomous programs with specific behavior. Agent stands for a computer system that is placed in an external environment that can interact with it, making the autonomous rational actions to achieve specific goals [6].

Fig. 2 schematically shows the multi-agent software architecture integrated into the portal of expert system.

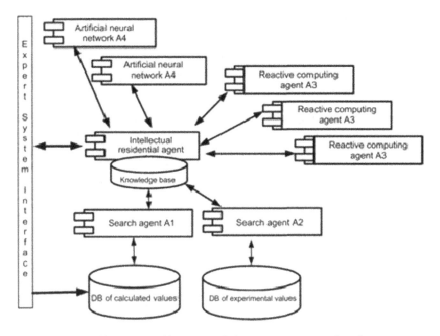

Fig. 2. Schematic multi-agent architecture of the expert system for the evaluation of bonds dissociation energy of organic molecular by the kinetic data of radical reactions

As can be seen from the Fig. 2, the expert system is served by the following agents: an agent of A1 type searches in the database of calculations. This agent offers its services, if in the calculated data of knowledge warehouse on bond dissociation energy some data are available; an agent of A2 type searches in the experimental data of knowledge warehouse. This agent offers its services, if in the experimental data on bond dissociation energy some data are available; an agent of A3 type estimates the bonds dissociation energies of molecule's on the basis of empirical model of radical reactions. This agent offers its services, if the vector of input parameters contains enough data for calculation. Evaluation of reactivity can be performed both in liquid phase and gas phase; an agent of A4 type uses the trained artificial neural network for estimating the bonds dissociation energy of molecular. This agent makes a decision on the provision of its services if its trained artificial neural network meets the specified input data. The agent has the opportunity to retrain its network offline.

Intellectual residential agent performs an analysis of input data, selects executing agents, examines the obtained result and returns it to the interface of the expert system.

The user through the expert system interface can save the obtained result in the database of calculated data by filling in the specific calculation form.

5   Conclusions

The development and publication on the Internet of subject-oriented science intelligence system based on the use of knowledge warehouses using the multi-agent technology will allow the scientific community to create distributed networks for the collection, storage, retrieval, mining, distribution and production of new knowledge in the specialized areas of research and technologies.

It should be noted, that the system represented in this paper, is the object independently functioning on the Internet: it is designed to evolve through the completion of its knowledge warehouses by users and experts.

The inclusion into such systems of the subsystem for distance learning of subject-oriented knowledge, greatly expands the range of their potential users (undergraduate and graduate students), which promotes the independent formation of their professional knowledge, and for the teaching staff of higher educational institutions this provides additional training materials and electronic resource guide.

Acknowledgments. The work is executed at support of the grant of the Russian Foundation for Basic Research 09-07-00297-a.

References

1. Hackathorn, R.: Science Intelligence. Can a Business Intelligence Approach Enable "Smart" Science? DM Review (2005), http://www.DMReview.com
2. Thierauf, R.J.: Effective Business Intelligence Systems. Quorum Books, Westport (2001)
3. Dong, Q., Yan, X., Chirico, R.D., Wilhoit, R.C., Frenkel, M.: Database Infrastructure to Support Knowledge Management in Physicochemical Data. In: 18th CODATA Conference, Canada, Montreal, p. 36 (2005)
4. Tumanov, V.E.: Data Warehousing and Data Mining in Thermochemistry of Free Radical Reactions. In: Fourth Winter Symposium on Chemometrics "Modern Methods of Data Analysis", pp. 28–29. IPCP Press, Russia (2005)
5. Tumanov, V.E.: Subject-Oriented Science Intelligence Systems. Information Technologies 5, 12–18 (2009)
6. Wooldridge, M., Jennings, N.: Intelligent Agents: Theory and Practice. Knowledge Engineering Review 10(2), 115–152 (1995)

An Innovative Contribution to Health Technology Assessment

Giovanni Improta[1], Maria Triassi[1], Guido Guizzi[2], Liberatina Carmela Santillo[2], Roberto Revetria[3], Alessandro Catania[3], and Lucia Cassettari[3]

[1] Dipartimento di Scienze Mediche Preventive, University of Naples "Federico II", Naples, Italy
{giovanni.improta,maria.triassi}@unina.it
[2] Dipartimento di Ingegneria dei Materiali e della Produzione, University of Naples "Federico II", Naples, Italy
{g.guizzi,santillo}@unina.it
[3] Dipartimento di Ingegneria Meccanica, University of Genoa, Genoa, Italy
{roberto.revetria,alessandro.catania,cassettari}@diptem.unige.it

Abstract. Healthcare is moving towards increased assistance needs with limited resources, both in economics terms, in personnel or space terms, leading to the usage of specific analysis for the acquisition, evaluation and assessment of medical technologies. The systematic evaluation of properties, effects or other impacts of a medical (or health) technology with a broad multidisciplinary approach is named Health Technology Assessment (HTA).This work presents an approach of a HTA protocol for the classification of hospitals or health facilities equipments, realized by combining the classic HTA concepts with hierarchic clustering techniques in a multidisciplinary analysis of requirements, cost, impact of logistics, technology associated risks.

Keywords: Health, Assessment, Facility, Multi-criteria approach, Decision making.

1   Introduction

The main purpose of HTA is to assist policymaking for technology in health care to achieve the most advantageous resource allocation, evaluating the efficacy and the efficiency of the diagnostic and therapeutic pathways as well as related risks and organizational models.

HTA consists in identifying an analytical methodology that allows the optimization of the product adoption/evaluation process, through a careful study of the effective needs of the users, of the available alternative technologies and the relative operational implications on the setup.

This kind of evaluation requires an interdisciplinary approach of "policy analysis", studying the aspects of safety, cost, benefits, effectiveness, and also include critical evaluations of the actual measures and improving the quality of life.

W. Ding et al. (Eds.): Modern Advances in Intelligent Systems and Tools, SCI 431, pp. 127–131.
springerlink.com   © Springer-Verlag Berlin Heidelberg 2012

HTA methodology implies to recognize the actual healthcare needs and evaluate how technologies may answer to those needs while considering the overall implication of their use including the associated risks.

It may address the direct and intended consequences of technologies as well as their indirect and unintended consequences.

HTA practices have become widespread and are increasingly present in health systems, so that more and more healthcare facilities monitor the global impact of their medical technologies.

The fundamental step of an HTA process can be summarized, as well as in a technology assessment, in some main steps that include:

- the *identification* of the assessment object/topic in order to clarify:
    - the problem addressed by the technology;
    - real clinical needs (needs assessment);
    - requirements or constraints the technology under investigation has to fit.
- the *evaluation* of the technology that for HTAs has to include:
    - the collection of key data in terms of general impact: technical, clinical, social, ethical as well as economical; it involves the comparison of different technologies according to criteria of quality, evaluating the clinical efficacy (benefits), safety, clinical outcomes, costs of the entire life cycle of technology;
    - the analysis of all the collected data and the technology rating;

This process, and its multidisciplinary evaluations, characterize (or it should do) all HTA processes.

- the *synthesis* phase includes:
    - the consolidation and synthesis of all the analysis in order to give a synthetic overview of the assessment results;
    - the production of recommendation on the applicability and adoption of the technology;

However, as in every dynamic process, monitoring the effectiveness of the assessment conclusion helps in refining methodologies and in assuring the correctness of the decision adopted.

2   Literature Review

During past decades, health care systems of industrialized countries have focussed on the problem of assuring health services to all citizens while reducing the allocation of economic resources.

To achieve both the subsistence of the essential health services and the reduction of sanitary costs, almost every state engaged in policies aimed at rationalizing the use of resources by acting on the efficiency of organizations in strengthening service delivery as well as introducing elements of competition between producers or prioritization of health care services to ensure to citizens through public funding. (Sackett, 1980; Battista and Hodge, 1989).

The need to evaluate the effectiveness of different diagnostic and therapeutic protocols and technologies compared to the suffering population and, at the same time, the need to a complete knowledge of the service delivery costs originated a multidisciplinary research area called "Health Technology Assessment". (Battista and Hodge, 1999).

Technical information needed by policymakers is frequently not available, or not in the right form. A policymaker cannot judge the merits or consequences of a technological program within a strictly technical context. He has to consider social, economic, and legal implications of any course of action.

3   Problem Solution

The protocol has been structured following a hierarchical assessment approach, similar to AHP (Saaty, 1980, Saaty, 1982, 1990), based on the definition of the goal to achieve, the criteria and evaluation parameters and their relative and global incidence in the overall decision.

A hierarchical breakdown of the problem in N different criteria (or cluster), which groups properties and attributes of alternatives, helps in a better evaluation of the problem itself.

For each cluster are then recognized properties or attributes (or cluster elements) in a variable number. It is worth mentioning that, in health environments, these properties and attributes are not always directly and objectively measurable (i.e. revenues versus degree of patient technology acceptance) and, in order to obtain a comprehensive and concise assessment reducing subjective bias, these are aggregated together in clusters.

The non-objective measurable parameters/attributes can also be quantified and then made comparable by using expert opinions expressed in linguistic variables and converted into numerical values (usually using the ordinal scale used in AHP and a pairwise comparison procedure with the aim of producing a square matrix, whose element aij indicates the relative importance of the element with respect to criteria A_j).

Synthetic assessment of the degree of importance of the single A_j with respect to the others   (weights) are calculated by normalizing the global importance of individual factors, i.e. the sum of each element of a row, with respect to the sum; it keeps unchanged the relationships between the factors and makes the sum of all weight obtained equal to 1, which is mathematically convenient in weighted sums.

Assuming gather experts evaluations so that $a_{ik} = a_{ij}*a_{jk}$ (i. e. assuming to know n-1 matrix elements and obtaining the remaining matrix elements from the properties of consistency and reciprocity) is not necessary to evaluate the AHP technique Consistency Index C.I. (CI) as it has supposed a perfect consistency in judgments (C.I. = 0).

Weights obtained are aggregated together with the hierarchical Saaty's composition principle, which allows a priority listing of alternatives to the goal.

In our case the final equipment classification is obtained by scoring the equipment based on the evaluated importance of the criteria and their properties which helps in correctly combining the specific characteristics/condition of the equipment under investigation.

Based on this principle, the overall score of the generic alternative A with respect to the goal may be expressed as:

$$C = \frac{\sum_{i=1}^{k} P_i \cdot V_i}{\sum_{i=1}^{k} P_i} \tag{1}$$

where:

k is the cluster numbers

P_i is the weight of cluster

V_i is the total score of equipment with respect to i-cluster

C is the total score of equipment

$$V_j = \frac{\sum_{j=1}^{n} p_j \cdot v_j}{\sum_{j=1}^{n} p_j} \tag{2}$$

where:

n is the element numbers

p_j is the weight of element j with respect to cluster

v_j is the score of element j

V_j is the total score of equipment for that criteria

Finally, the process ends with a classification of the equipment based on its specific score; in particular, since we hypothesized four different alternatives, classification is achieved choosing three different thresholds and comparing the obtained equipment score with those values. In case of partial evaluation (that is the evaluation based only on some cluster) the sum will include only the aspect under investigation.

4   Conclusions

Technology assessment cannot replace the clinical governance decision makers as these topics are often related to variables dependent on their sensitivity; however, HTA certainly improve management processes through a more effective use of information and knowledge available.

HTA leads to a wider risk analysis and a better health needs assessment, it makes possible an extensive knowledge of the technology characteristics, its effects on individuals health, its economic and/or organizational impact and may allow:

-   improved selection processes: for the selection of technologies to adopt through an explicit comparison between the "needs" (health needs, resources available);
-   efficient management of procurement processes, since a better understanding of the overall characteristics of the technology can enhance negotiation skills in dealing with suppliers;

– the preparation of all the organizational, professional and financial resources necessary for effective and efficient use of technology in order to increase the level of performance provided.

Generally, HTAs are mainly related to technology or equipment purchase; results are presented in the form of reports or indicators to help decision makers in their conclusions.

However, in our knowledge few of them have been dedicated to a classification of the overall state of hospital equipments especially with relocation/donation purposes.

The proposed methodology, based on the requirements and constraints often suggested by decision makers themselves, provides an indicator (a numerical value) through which the equipment may be classified; using the algorithm all the information associated with the assessment are synthesized to allow managers in easily getting an overall picture of capital equipment state and usage implications in the hospital facility.

References

1. Bates, D.W., Gawande, A.A.: Improving safety with information technology. New Engl. J. Med. 348(25), 2526–2534 (2003)
2. Battista, R.: Innovation and diffusion of health-related technologies. A conceptual framework. Int. Journal Technol. Asses. Health Care 5(2), 227–248 (1989)
3. Battista, R., Hodge, M.J.: The Evolving Paradigm of Health Tecnology Assessment: Reflections for the Millennium. CMAJ 160(10), 1464–1467 (1999)
4. Granados, A., Sampietro Colom, L., et al.: Health Technology Assessment in Spain. International Journal of Technology Assessment in health Care 16(2), 532–559 (2000)
5. Jorgensen, T., Hvenegaard, A., et al.: Health Technology Assessment in Denmark. International Journal of Technology Assessment in health Care 16(2), 347–381 (2000)
6. Kahveci, R.: Analysis of strengts, weakness, threats and opportunities in development of an HTA program in Turkey. Handb Health Technol Assess Int. (2006)
7. Kristensen, F.B., Horder, M., et al.: Health Technology Assessment handbook. Danish Institute for HTA (2001)
8. Perleth, M., Busse, R.: Health Technology Assessment in Germany. International Journal of Technology Assessment in health Care 16(2), 412–428 (2000)
9. Sackett, D.L.: Evaluation of health services. In: Last, J.M. (ed.) Health and Preventive Medicine, pp. 1800–1823. Appleton-Century Crofts, New York (1980)
10. Sackett, D.L., Rosenberg, W.M.C., Gray, J.A., Haynes, B., Richardson, S.: Evidence based medicine: what it is and what it isn't. British Med. J. 372(13) (1996)
11. Saaty, T.L.: The Analytic Hierarchy Process. McGraw Hill, New York (1980)
12. Saaty, T.L.: How to make a decision: The Analytic Hierarchy Process. European Journal of Operational Research 48(1), 9–26 (1990)
13. Sorkin, A.L.: Health economics: an introduction. Lexington Books, Lexington (1975)
14. Weisbrod, B.A.: Economics of public health. University of Pennsylvania Press, Philadelphia (1961)

Machine Learning

Bloom's Taxonomy–Based Classification for Item Bank Questions Using Support Vector Machines

Anwar Ali Yahya, Zakaria Toukal, and Addin Osman

Faculty of Computer Science and Information Systems, Najran University, Najran, Kingdom of Saudi Arabia
{aaesmail,zstoukal,aomaddin}@nu.edu.sa

Abstract. This paper investigates the effectiveness of support vector machines for the classification of item bank question into Bloom's taxonomy cognitive levels. In doing so, a dataset of pre-classified questions has been collected. Each question has been processed through removal of punctuations, tokenization, stemming, term weighting, and length normalization. Using this dataset, the performance of support vector machines has been evaluated considering the effect of term frequency and stopwords removal. The results show a satisfactory performance of support vector machines, which declines as the frequency of the terms used to represent question increases. The best performance is obtained when term frequency is greater than or equal to two. Moreover, the results show that the removal of stopwords does not improve the performance significantly.

1 Introduction

An item bank is an electronic repository (database) of test items which are periodically collected at each test or exam time over years and stored along with information pertaining to those items. It is a valuable resource for E-assessment applications. It can be used to design effective tests by allowing a unique subset of questions to be chosen for each test or student where specific or personalized skills and levels of competence need to be examined. In the item bank, each item consists of question, answer, and metadata. An important type of item's metadata is Bloom's taxonomy cognitive levels (BTCLs). In the field of education, Bloom's taxonomy is an essential concept which was developed by Benjamin Bloom [1] in his efforts to classify the thinking behaviours. He identified three domains: cognitive, affective and psychomotor. Under the cognitive domain, Bloom identified six different levels of learning as follow:-

- Knowledge: Memorization, recognition, and recall of information.
- Comprehension: Organization of ideas, interpretation of information.
- Application: Problem solving, use of particulars, and principles.
- Analysis: Finding the underlying organization.
- Synthesis: Combination of ideas to form something new.
- Evaluation: Making judgments about issues.

W. Ding et al. (Eds.): Modern Advances in Intelligent Systems and Tools, SCI 431, pp. 135–140.

It goes without saying that the manual classification of item bank questions into BTCLs is time consuming, tedious, and prone to mistake. Therefore, this paper presents an original work on the use of Support Vector Machines (SVMs) to automate this task.

2   Related Works

It is obvious that the classification of item question into BTCLs can be casted as text classification problem. Since its first appearance, text classification has been used in a good number of applications either explicitly as the main technology or implicitly, as a supportive technology. In the field of E-learning, text classification has been used in a number of applications [4, 5]. For the particular application of questions classification several works have been reported. In [10], back-propagation neural network is proposed to classify question into three difficult levels; easy, medium, and hard. Another work [2] focuses on the classification of a specific type of questions, called open-ended questions using SVMs. An interesting work on using SVMs for question classification is presented in [7], in which an adaptable learning assistant tool for managing item bank is presented.

3   SVMs for Item Question Classification

SVM is a supervised machine learning algorithm that is trained to separate between two sets of data using training examples of both sets. SVM approach was introduced to text classification by Joachims [3] and subsequently used by many researchers [6, 9]. Joachims argued that SVMs are very suited for text classification due to the inherent characteristics of text (i.e, high dimensional input space, few irrelevant features, the sparseness of documents vectors, and the linear separability of most text classification problems). In addition to that, the ability of SVMs to generalize well in high dimensional feature spaces, such as text classification, eliminates the need for feature selection.

Using SVMs for text classification system requires three main steps: text representation, SVMs classifiers construction, and SVMs classifiers evaluation. In the following subsection a detailed description of these steps, in the context of item bank questions classification into BTCLs, is given.

3.1   Item Question Representation

Each question q_j is represented as a vector of term weights $<w_{1j}, ..., w_{Tj}>$, where T is the set of terms that occur at least once in at least one question , and $0 \leq w_{kj} \leq 1$ represents, how much term t_k contributes to the semantics of question q_j. The term weight can be a binary weights or non-binary. For the current application of SVMs, non-binary weights (standard *tfidf* function) is used, which is defined as follows

$$tfidf\ (t_k, q_j) = \#(t_k, q_j).\log \frac{|Tr|}{\#Tr(t_k)} \tag{1}$$

Where $|Tr|$ is the number of questions in the training set, $\#(t_k, q_j)$ is the number of times t_k occurs in q_j , and $\#Tr(t_k)$ is the question frequency of term t_k, that is, the number of questions in which t_k occurs. In order to apply the above representation a preprocessing of question is applied which includes:

- Reducing question text to lower case characters.
- Punctuation removal: All types of punctuations are removed from the question.
- Tokenization: A token is a maximal sequence of nonblank characters.
- Stemming: The tokens were stemmed with Porter stemmer [8].

After the preprocessing of a question text, term weighting is computed as in Eq. 1, and length normalization is applied as follow:

$$w_{q_j}^{''}(t_k) = \frac{w_q(t_k)}{\sqrt{\sum_l w_{q_j}(t_l) \times w_{q_j}(t_l)}} \tag{2}$$

3.2   SVM Classifiers Construction

In this step, a single SVM classifier is trained for each BTCLs class using a subset of dataset, called training set. This step can be accomplished using one of the currently available SVMs tools. In this work SVM-Light package, version 6.02 has been used. It is freely available from http://svmlight.joachims.org/.

3.3   SVMs Classifiers Evaluation

The effectiveness of the SVMs on a single class of BTCLs can be evaluated using several measures. The computation of these measures depends essentially on a contingency table obtained from the classification of the testing set of each BTCLs class. The contingency table of a given BTCLs class contains the following values

- A: The number of questions a system correctly assigns to the class.
- B: The number of questions a system incorrectly assigns to the class.
- C: The number of questions belonging to the class but the system does not assign them to that class.
- D: the number of questions a system correctly doesn't assign to the class.

The following are the common measures of a single SVMs classifier.

- Precision:- It is defined as follow:

$$P = \frac{A}{A + B} \tag{3}$$

- Recall: It is defined as follow:

$$R = \frac{A}{A + C} \tag{4}$$

- F_β measure: The harmonic mean of recall and precision, for $\beta=1.0$, as follow:

$$F_{1.0} = \frac{2\,RP}{R + P} \tag{5}$$

- Accuracy: The accuracy of a classifier is defined as follow:

$$Acc = \frac{A + D}{A + B + C + D} \tag{6}$$

Additionally, SVMs effectiveness across a set of classes can be measured by the macroaverage (unweighted mean of effectiveness across all classes) and the microaverage (effectiveness computed from the sum of per-class contingency tables).

4   Experimental Results

This section presents the results obtained from two series of experiments. The first series was devoted to evaluate the performance of SVM under different values of term frequency. The second series investigates the effect of stopwords removal. In order to conduct these experiments, a dataset of 600 questions has been collected from a several item bank available on the Web and classified manually by a pedagogical expert. The questions are distributed evenly among classes (i.e. has 100 examples for each). The dataset has been processed as described in section 4.1 and divided into training set (70% of the dataset) and testing set (30% of the dataset).

4.1   SVMs Performance and Term Frequency

The results of the first series of experiment are shown in Table 1.

Table 1. SVMs Performance and Term Frequency

	$TF \geq 1$		$TF \geq 2$		$TF \geq 3$		$TF \geq 4$		$TF \geq 5$	
	Acc.	F	Acc.	F	Acc.	F	Acc.	F	Acc.	F
Mac-Avrg	0.907	0.622	0.923	0.711	0.919	0.699	0.916	0.622	0.909	0.665
Mic-Arag	0.907	0.638	0.923	0.717	0.919	0.707	0.916	0.638	0.909	0.669
	$TF \geq 6$		$TF \geq 7$		$TF \geq 8$		$TF \geq 9$		$TF \geq 10$	
	Acc.	F	Acc.	F	Acc.	F	Acc.	F	Acc.	F
Mac-Avrg	0.904	0.628	0.890	0.560	0.873	0.460	0.855	0.370	0.852	0.314
Mic-Arag	0.904	0.639	0.890	0.577	0.873	0.487	0.855	0.389	0.852	0.344

The results show that as the term frequency increases, the performance of SVMs declines in all aspect. Another conclusion that can be drawn from the results is that the best performance of SVMs can be obtained when the $TF \geq 2$.

4.2   SVMs Performance and Stopwords Removal

Table 2 shows the results of the second series of experiments.

Table 2. SVM Performance with Stopwords Removal

	TF ≥ 1		TF ≥ 2		TF ≥ 3		TF ≥ 4		TF ≥ 5	
	Acc.	*F*	*Acc.*	*F*	*Acc.*	*F*	*Acc.*	*F*	*Acc.*	*F*
Mac-Avrg	0.907	0.636	0.920	0.704	0.915	0.665	0.917	0.700	0.906	0.651
Mic-Avrg	0.907	0.648	0.920	0.711	0.915	0.683	0.917	0.706	0.906	0.658
	TF ≥ 6		TF ≥ 7		TF ≥ 8		TF ≥ 9		TF ≥ 10	
	Acc.	*F*	*Acc.*	*F*	*Acc.*	*F*	*Acc.*	*F*	*Acc.*	*F*
Mac-Avrg	0.890	0.569	0.882	0.523	0.867	0.409	0.857	0.371	0.848	0.302
Mic-Avrg	0.890	0.582	0.882	0.551	0.867	0.450	0.857	0.408	0.848	0.339

The results support the previously drawn conclusion on the performance of SVMs with different values of term frequency. However, a comparison of SVMs performance in both cases shows insignificant difference, as shown in Fig. 1.

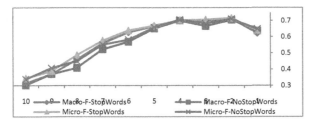

Fig. 1. SVM Macro (Micro)-Average F Performance without (with) Stopwords Removal

5   Conclusion

This paper introduces an automatic classification approach for item bank questions using SVMs. The obtained results show a satisfactory performance of SVMs, with respect to accuracy and F-measure. The best performance of SVMs is obtained when terms with frequencies ≥ two are used for representation. They also show that the removal of stopwords does not improve SVMs performance significantly.

Acknowledgments. This work is supported by the Scientific Research Deanship in Najran University, Kingdom of Saudi Arabia under research project number NU 21/11.

References

1. Bloom, B.S.: Taxonomy of educational objectives. Handbook I. The Cognitive Domain. David McKay Co. Inc., New York (1956)
2. Bullington, J., Endres, I., Rahman, M.: Open ended question classification using support vector machines. In: MAICS (2007)

3. Joachims, T.: Text Categorization with Support Vector Machines: Learning with Many Relevant Features. In: Nédellec, C., Rouveirol, C. (eds.) ECML 1998. LNCS, vol. 1398, pp. 137–142. Springer, Heidelberg (1998)
4. Jung-Lung, H.: A text mining approach for formative assessment in e-learning environment. Dissertation, National Sun Yat-sen University, Taiwan (2008)
5. Karahoca, A., Karahoca, D., Ince, F.I., Gökçeli, R., Aydin, N., Güngör, A.: Intelligent question classification for e-learning by ANFIS. In: E-learning Conference 2007, pp. 156–159 (2007)
6. Klinkenberg, R., Joachims, T.: Detecting concept drift with support vector machines. In: Proc. of the 17th Int. Conf. on Machine Learning, ICML 2000, Stanford, CA, pp. 487–494 (2000)
7. Nuntiyagul, A., Naruedomkul, K., Cercone, N., Wongsawang, D.: Adaptable learning assistant for item bank management. Computers & Education 50, 357–370 (2008)
8. Porter, M.F.: An algorithm for suffix stripping. Program 14(3), 130–137 (1980)
9. Taira, H., Haruno, M.: Feature selection in SVM text categorization. In: Proc. of the 16th Conf. of American Association for Artificial Intelligence, AAAI 1999, Orlando, FL, pp. 480–486 (1999)
10. Wiggins, G.: Educative assessment designing assessments to inform and improve student performance. Jossey-Bass Inc., California (2003)

Analyzing Question Attractiveness in Community Question Answering

Xiaojun Quan and Liu Wenyin

Department of Computer Science, City University of Hong Kong
Tat Chee Avenue, Kowloon, Hong Kong SAR
xiaoquan@student.cityu.edu.hk, csliuwy@cityu.edu.hk

Abstract. There have been increasing interests in Community Question Answering (CQA) recently. CQA websites such as Yahoo! Answers and Baidu Knows are increasingly popular, attracting tens of thousands of users to submit questions and answers every day. However, we find that there is a gap in the study of what kinds of questions are more likely to attract answers, which makes sense for users in practice when asking questions in the CQA systems. In this paper, we investigate the factors that may affect users' questions to attract answers. We also introduce a general framework to predict how many replies a question is expected to receive. Evaluation results of the framework on real data prove its effectiveness.

1 Introduction

There has been a recent trend towards referring to a community question answering (CQA) portal when users encounter problems, especially when search engines return very poor results for their questions. CQA [1] has been drawing more and more attention from both academia and industry. There are many popular CQA systems attracting tens of thousands of users every day. Yahoo! Answers[1] is one of the most successful CQA websites which has gained wide popularity. Latest statistics [2] show that Yahoo! Answers has received more than 1 billion questions and answers from across the world since it is launched. It counts more than 179 million users across 26 countries or regions, and around 15 million users visit the website every day. Naver Knowledge-iN[2], a Korean CQA portal, has accumulated a human-generated database of 70 million entries. Baidu Knows[3] is a leading Chinese CQA service provider. From the data on the landing page of this website more than 164 million questions have been resolved in this CQA site until November 2011.

For newcomers in the CQA website, as well as unskillfully regular users, they often have to wait a long time before they can obtain the best answers. The reasons are multifold. For example, they are asking uncommon and weird questions and

---

[1] http://answers.yahoo.com
[2] http://kin.naver.com
[3] http://zhidao.baidu.com

W. Ding et al. (Eds.): Modern Advances in Intelligent Systems and Tools, SCI 431, pp. 141–146.
springerlink.com   © Springer-Verlag Berlin Heidelberg 2012

thus other users find the questions unintelligible or difficult to provide answers. Other reasons, such as the scale of online users and reward points for their questions, may also affect askers to attract many answers. Given that the problem is very common to users and there have been a lot of existing researches in CQA [3][4][5][6][7], we find relatively little is known about details of what factors affect users' satisfaction in CQA websites. In this paper, we investigate the several non-text factors of questions, including question length, reward points, the existent answers and question submission time. Based on the above factors, we propose a general prediction framework of how many answers a question is expected to receive. Experimental results witness the effectiveness of this prediction model.

2   Factors Analysis in Community Question Answering

In this section, we first introduce the data collections used in this study and then explore the factors that influence questions being answered.

2.1   Collections

Liu et al. [7] used a large collection crawled from Yahoo! Answers in their investigation of predicting users' satisfaction in CQA, which is designated as the first collection in this study. The collection consists of 216,563 English questions from 100 categories of Yahoo! Answers. Statistics on this collection shows that the average number of words (question length) of a question before and after removing stop words is about 9.75 and 4.70, respectively. There are totally 1,981,830 answers. The other collection is crawled from Baidu Knows website. We obtain a total of 2,049,334 Chinese questions and 6,497,523 answers to these questions, which contains an average of 3.17 answers to each question. The question length in this collection varies from 1 to 623, with an average length of 57.6. There is reward information (reward points) in this collection.

2.2   Question Length

Long questions are usually more difficult to read and comprehend than short ones. Answerers may avoid answering long questions in order to save time and effort. We give a comprehensive study on all the questions in the collections. We re-organize the questions into different categories according to their lengths, and therefore questions with the same length will be classified into one category. We calculate the average number of answers to a question in each category and plot the histograms in Figure 1. We can observe from the histograms that, for both collections, the questions with average length from 5 to 10 tend to receive more answers than others. It implies that questions with length within this scope are much easier to get answers.

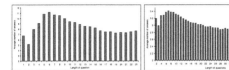
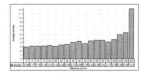

Fig. 1. Histograms of average number of answers for each question with different lengths on Yahoo! Answers (left) and Baidu Knows (right).

Fig. 2. Average number of answers with different reward points

2.3   Reward Points

Some CQA websites, such as Baidu Knows, encourage users to ask questions with incentives, such as reward points. User whose answer has been chosen as the "best answer" will get the incentives. Reward points in Baidu Knows can be used to spend for other services in the website. Therefore, users will be encouraged to provide high-quality answers to earn the reward points. The Baidu Knows collection is chosen for this test because many questions in this collection contain reward points information. The questions are organized into different categories according to their reward points (maximum 200) offered by question authors. We then calculate the average number of answers of a question in each point category. Results are plotted as histograms shown in Figure 2. It is easy to find that, with the reward point increases, the average number of answers is also increasing. However, compared to questions without rewards, there is not significant increase in the average number of answers for questions whose points are less than 150.

2.4   Impact of Existent Answers

Intuitively, users are unwilling to reply questions which have already attracted a lot of answers. The rationale of this assumption is that more answers of a question indicates higher possibility that the question asker will be satisfied with existing answers, and there is less chance for new answers to be helpful. We examine the distribution of questions over different numbers of answers by first recording the number of questions with at least one answer in Yahoo! Answers and Baidu Knows collections respectively, and then continue until two and more answers. The results are given in the form of curves in Figure 3. We can find that there are large amounts of questions with at least one answer. But, the scale of questions with at least two or more answers decreases dramatically. Among the questions having received the first answer, there are about 25 percent of them will not get another answer. After the questions have got many answers, such as 40 for Yahoo! Answers and 10 for Baidu Knows, few of them can receive another reply.

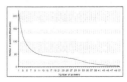
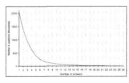

Fig. 3. The distribution of questions over different numbers of answers. Yahoo! Answers (left) and Baidu Knows (right).

2.5   Post Time of Questions

The time that a question is posted corresponds to the number of online users in the CQA website. Note that even though a question is published in a perfect time, there are still a lot of follow-up answers after several hours or even a few days. For this reason, we consider only the time of the first answer and the second answer of a question in this study.

We evenly divide one day into 24 time periods and then organize all the questions into the 24 categories based on the time periods they are raised. After removing the questions without time information, we obtain a total of 215,383 questions with at least one answer from Yahoo! Answers collection and 717,431 questions from Baidu Knows collection. For the case of with at least two answers, we get in total 156,278 questions from Yahoo! Answers collection, and 411,747 questions from Baidu Knows collection. Next, the average time of a question to get the first/second answer in each category is computed to measure how efficient in certain time period the questions are being answered. Noted that the time of the first and the second answers still varied dramatically, ranging from several seconds to more than one month. Here, we employ the "five-number-summary" strategy [8] in descriptive statistics for each category, and computed the average time using fifty percent of the data. Specifically, the data in each category is sorted in descending order, and only the middle 50% data is used to calculate an average result. The overall results on two collections are shown in Figure 4. Generally, users in Yahoo! Answer are expected to get their first and second answers very soon in from 11:00 to 24:00, as shown in (a) and (c) of Figure 4. For Baidu Knows, there is a relatively clear pattern, shown in figure (b) and (d), that users will soon receive answers for their questions in the range of 7:00 to 18:00.

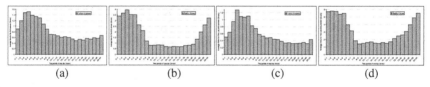

(a)   (b)   (c)   (d)

Fig. 4. The average time of for the first/second answer's appearance within different time periods. (a) The first answer in Yahoo! Answers collection; (b) The first answer in Baidu Knows collection; (c) The second answer in Yahoo! Answers collection; (d) The second answer in Baidu Knows collection.

3   Attractive Questions Prediction

In this section, we aim to predict whether a question will be followed by many users after its submission based on the above factors. Of the four factors, the factor "existent answers" is not available when a question is newly posted so we just ignore it in this part. In addition, apart from the above non-text features, the content of questions is also taken into account.

3.1   Methodology

Intuitively, questions with similar features are likely to get similar numbers of answers. We employ a k-nearest neighbors strategy to find similar questions to the new question in features and predict the number of answers to the question, with the following steps. First of all, for each test question, the k-nearest neighbors of this question are firstly found from training question set. The nearest neighbors are detected based on the distance of two questions, q_i and q_j, which can be calculated as $D(q_i, q_j)$ $= \beta_1 DL_{ij} + \beta_2 DR_{ij} + \beta_3 DT_{ij} + \beta_4 DS_{ij}$. Here DL_{ij}, DR_{ij} and DT_{ij} denote the length, reward points and time distances of question q_i and q_j, which are calculated by $|x-y|$, in which x and y represent the same aspects (such as length, reward points and time) of two questions. X and y are represented as described in Table 1. DS_{ij} is the textual distance between q_i and q_j which is calculated as the reciprocal of the semantic similarity of the two questions. The similarity of q_i and q_j is obtained based the cosine metric with *tf-idf* as the term weighting scheme. β_1, β_2, β_3, and β_4 are parameters subjecting to $\beta_1 + \beta_2 + \beta_3 + \beta_4 = 1$. Next, the number of answers of this test question is computed by averaging the number of answers of its neighbor questions as:

$$N_t = \frac{1}{k}\sum_{i=1}^{k} N_i ,$$

where N_t is the number of answers for the test question, and N_i is the count of answers of its i-th neighbor.

Table 1. Non-text features used for prediction

Features	Value	Description
Question length	$n/100$	n: the number of words of a question; 100: the maximum length
Reward points	$r/200$	r: the reward points; 200: the maximum reward points.
Time	$t/24$	t: the time that a question was posted, represented with real numbers and hour as the unit.

3.2   Experimental Setting and Result

The optimal parameters β_1, β_2, β_3 and β_4 are estimated with an enumerating strategy. The performance is measured in terms of *precision*, which is calculated as the number of correctly predicted test cases divided by the total number of test cases. The parameter k is fixed to 30 according to empirical test. We set a reasonable error range, σ, for the prediction, which means that if the predicted number of answers for a question is within this range compared to the actual situation, we treat this prediction as correct. For example, if a test question has received 5 answers actually, and the predicted number for it is 4, this prediction is correct subjecting to σ is 1. However, if the predicted number is 3, the prediction is failed.

We construct a larger training dataset from Baidu Knows collection by randomly selecting 100,000 questions. The k-nearest neighbors strategy is then performed on the dataset with 10-fold cross-validation based on the above parameters setting. The final result of this test is shown in Table 2. Note that since we have not discovered any similar work, there is not available baseline can be used for comparison. Result in the table shows that the prediction method is effective for the prediction task.

Table 2. Prediction precision on the first dataset

σ	1	2	3
precision	0.5331	0.7879	0.8858

4   Conclusions

This study investigates four non-text factors related to questions. We have the following conclusions. (a) Short questions are more likely to attract answers. (b) Users are expected to receive more answers with the reward points/incentives increase. (c) Users are more willing to reply questions in the early submission stage of question. (d) The submission timing of questions is also important. Based on the above analysis, we propose a framework to predict how many answers a question is possible to receive and experimental results prove its effectiveness.

References

1. Wenyin, L., Hao, T.Y., Chen, W., Feng, M.: A web-based platform for user-interactive question-answering. World Wide Web: Internet and Web Information Systems 12(2), 107–124 (2009)
2. Yahoo! Answers blog, `http://yanswersblog.com/index.php/archives/2009/10/05/did-you-know`
3. Jurczyk, P., Agichtein, E.: Discovering authorities in question answer communities by using link analysis. In: Proc. of CIKM (2007)
4. Agichtein, E., Castillo, C., Donato, D., Gionis, A., Mishne, G.: Finding high-quality content in social media. In: Proceedings of the International Conference on Web Search and Web Data Mining, WSDM 2008 (2008)
5. Harper, F.M., Raban, D., Rafaeli, S., Konstan, J.A.: Predictors of answer quality in online Q&A sites. In: Proceeding of the Twenty-Sixth Annual SIGCHI Conference on Human Factors in Computing Systems (2008)
6. Bian, J., Liu, Y., Zhou, D., Agichtein, E., Zha, H.: Learning to recognize reliable users and content in social media with coupled mutual reinforcement. In: Proceedings of the 18th International Conference on World Wide Web, WWW 2009 (2009)
7. Liu, Y., Bian, J., Agichtein, E.: Predicting information seeker satisfaction in community question answering. In: Proc. of SIGIR Conference on Research and Development in Information Retrieval, SIGIR 2008 (2008)
8. `http://en.wikipedia.org/wiki/Five-number_summary`

An Improvement for Naive Bayes Text Classification Applied to Online Imbalanced Crowdsourced Corpuses

Robin M.E. Swezey, Shun Shiramatsu, Tadachika Ozono,
and Toramatsu Shintani

Graduate School of Engineering, Nagoya Institute of Technology
Gokiso-cho, Showa-ku, Nagoya, Aichi, 466-8555 Japan
{robin,siramatu,ozono,tora}@toralab.ics.nitech.ac.jp

Abstract. In order to be able to use the advantage of public corpuses such as Wikipedia to address problems of classification by hierarchically structured topics with a large amount of classes, we propose an improvement of Naive Bayes based text classification algorithms which we call Semantically-Aware Hierarchical Balancing. SAHB addresses two issues in that specific use case with real-world applications, namely the large amount of topic labels to classify against, and the lack of balance in the hierarchy of the corpuses. This meta-algorithm performs with better accuracy and log-time complexity than straightforward naive bayes text classification methods or specific document weighing techniques, whilst taking equivalent time to train, which makes it more efficient, and also scalable to process and classify big data.

1 Introduction

The algorithm described in this paper originally stemmed from a broader research project, of which the aim is to provide an intelligent platform for assessing concern from online users and increase public involvement of the Japanese population about social issues. Such a platform's end-user perspective consists in various interfaces which present coherent data out of fuzzy and random Web data which is neither annotated, nor structured or consistent. To this end, the data has to be modeled adequately, which begins with labeling the various contents that we need to use (articles, tweets, etc).

In this paper, we focus mainly on our general approach to classify contents more efficiently and quickly according to geography by using open crowdsourced corpuses that can be found online and out of which datasets can be easily built and rebuilt. We discuss the research background, then present the approach and early validation experiments that show promising results. We then discuss further research directions and conclude.

2 Research Background

In order to use advantages of crowdsourced corpuses, one has to take account of several challenges. The corpus can be highly imbalanced, as such is the case for

W. Ding et al. (Eds.): Modern Advances in Intelligent Systems and Tools, SCI 431, pp. 147–152.
springerlink.com   © Springer-Verlag Berlin Heidelberg 2012

the Japanese Wikipedia corpus once labeled with confidence to geography: for each class c of a possible set of classes C such as topical classes or geographical classes, some classes contain a much bigger amount of documents than the others. There can also be a lack of data that can be neglected in the case of the biggest Wikipedia corpuses, but not in others. In smaller Wikipedia corpuses, where the language is not popular, the advantages stated above may not apply.

As a straightforward approach, we use a Transformed Weight-normalized Complementary Naive Bayes classifier (TWCNB) [4] which has a natural better efficiency than Naive Bayes (NB), as well as a short training time compared to SVM, while retaining close performance. In our preliminary experiments for geographical classification, TWCNB shows better efficiency than NB, so it will be the straightforward algorithm that we use in our sample problem. Details on how TWCNB functions can be found in [4].

As shown in our experiments in Section 4, TWCNB cannot be used straightforwardly in the case of geographical classification using the Japanese Wikipedia corpus. The bias created by class imbalance being too high, it needs to be supervised by a meta-algorithm. Such meta-algorithms for improving classifiers already exist, like AdaBoost. However, if the amount of classes is to become relatively large when compared to classical examples of multi-class Bayesian classification problems (20NewsGroups, Reuteurs, etc), even with AdaBoost improvement [2] the testing algorithm has a linear time complexity of $O(n)$. Hence, any straightforward testing algorithm becomes non-optimal in the case of nation-scale geographical classification of text, when the order of magnitude of the number of classes is 100 times higher.

Moreover, when it comes to accuracy, even with AdaBoost improvement, it is intuitive to take advantage of the tree-structure of geographical topic nodes in order to improve efficiency. Hierarchical approaches are more intuitive and appropriate in such cases, and exist for text classification with SVMs [1] as well as for known problems [6]. Here, we approach the problem of crowdsourced corpuses and use clustering of existing hierarchy nodes together to counter bias when the hierarchy of classes is known. Oversampling and undersampling [3] also are known methods for alleviating class imbalance problems, however they can also affect the class prior calculation and other variables in Bayesian classification learning.

3   System and Algorithm

We propose our method, which we call Semantically-Aware Hierarchical Balancing (SAHB). SAHB is a variant of resampling which creates upper-level class nodes by clustering semantically close sets of documents from lower-layer classes in a higher-layer classification problem and keeps the variance in the class size at its lowest possible, hence hierarchically balanced.

In training, the meta-algorithm takes all the predefined classes and groups semantically close classes under a same parent node. In testing, each node is a class (label) and a test for classification is applied at each depth level of the tree. This method is *semantically aware* because child nodes are clustered under a parent node which has semantical relation to it. It is *hierarchically balanced* because it takes account of the overall amount of classes for optimal classification. Bias in the class size affects the calculation of TfIdf weights, thus we cluster classes in a tree hierarchy to counter this bias.

In our classification example, at depth level n of the classification tree, the set of classes C_n is made of Tokyo, which holds a very high number of documents, and other prefectures from the Kanto region.

$$C_n = \{Tokyo, Chiba, Ibaraki, Saitama, ...\}$$

Because of high bias, a lot of documents are likely to be misclassified in Tokyo when testing. We indeed observed a quite heavy column for Tokyo in the confusion matrix when testing the classifier *flatly* (as opposed to hierarchically), i.e when SAHB was not applied. Hence, it is better to cluster classes geographically close to Tokyo against the Tokyo class itself to reduce the variance, with a set such as

$$C_n = \{Tokyo, \{Chiba, Ibaraki, Saitama, ...\}\}$$

The set at depth level $n+1$, if the document was not classified in Tokyo, is then

$$C_{n+1} = \{Chiba, Ibaraki, Saitama, ...\}$$

As a testing algorithm, SAHB also shows better time complexity than straightforward TWCNB (or AdaBoost-improved TWCNB) does. The time complexity is

$$\beta * log_\beta(n)$$

where β is the average branching factor of the class tree, which is considered stable and not a function of depth, and n the number of leaf classes to be tested against, which corresponds to multiplying β testing iterations by the β-ary search through the tree. Hence, it corresponds to a

$$O(log(n))$$

which is very suitable if n grows very large, for example if we need to classify not only against prefectures, but also cities and even districts.

4   Evaluation

In the following experiments, we describe the algorithm that we used with:

(Name of Meta-Algorithm)-(Depth of Class Tree)LH-(Name of Sub-Algorithm)

LH stands for *Layer Hierarchical*. We call *Flat* the case where there is no depth in the class tree, i.e when classification is done by testing the document against

Table 1. Performance of Flat TWCNB on Prefectures

Classes	47 prefectures
Correctly Classified Instances	69944 (70.250%)
Incorrectly Classified Instances	35074 (29.750%)
Total Classified Instances	11173
Avg. Time to Classify one Instance	12,4 ms

all label nodes at the same time. When no meta-algorithm is described, we simply used the natural hierarchy given by Wikipedia, without any clustering against bias.

The dataset for training and testing was built from the Japanese Wikipedia corpus, by adding each article as a document for the prefecture/city/district class it relates to. The Japanese Wikipedia constitutes a fairly rich corpus, but it also has the flaw of high imbalance, with majority classes such as the Tokyo label.

4.1   Validation Experiments

First, we conducted straightforward experiments to assess the existence of a problem related to class imbalance. Results are shown in Table 1.

In a per-city hierarchical classification experiment, we used the natural hierarchy given by Wikipedia, without any clustering against bias. This was also done in a previous paper [5] in which classification still needed improvement. For 2LH-TWCNB on Japanese cities, we obtained 66.602% accuracy with 38ms of average time to classify an instance, against respectively 42.086% accuracy and 522ms processing time in Flat-TWCNB.

As can be expected, a hierarchical classifier ensemble method outperforms the flat classifier in validation, notably in processing time performance, although its accuracy needs improvements. This is also important when determining confidence faster in the real-world testing of contents: if there is no confidence for the classifier in choosing between classes over the initial cluster of classes (regions which are clusters of prefectures), it is very unlikely that it will gain more confidence at a lower level of the class tree. Thus it also improves time complexity, speed and accuracy when testing confidence for analysis of the geographic property of contents, namely testing if contents are related to geography or not at all. Therefore, the discarding of numerous contents which are not geographically-relevant will also be significantly faster.

Finally, in a per-prefecture classification experiment, we use SAHB to cluster semantically-close label classes in cluster nodes, which are region nodes created for reducing bias. Nodes are built in the Japanese Wikipedia geographical classification example as described in Table 3.

We witness, as for hierarchical classification, a gain in processing time, but here also more notably a significant gain in accuracy with SAHB on Table 4 when compared to simple hierarchical classification on Table 2.

Table 2. Performance of 3LH-TWCNB on Prefectures

Training Classes	55 classes: 47 prefectures child nodes 8 region cluster nodes
Test Classes	47 prefectures
Correctly Classified Instances	10565 (75.360%)
Incorrectly Classified Instances	608 (24.640%)
Total Classified Instances	11173
Avg. Time to Classify one Instance	3,9 ms

Table 3. Cluster nodes built with SAHB in the prefecture experiment

First-level Cluster Node	Child Nodes
Hokkaido-Tohoku	Hokkaido, Tohoku
Kanto	Tokyo, KantoSub
Chubu	Aichi, Fukui, Gifu, Ishikawa, Nagano, Niigata, Shizuoka, Toyama, Yamanashi
Kansai	Osaka, KansaiSub
Chugoku-Shikoku-Kyushu	Hiroshima, Okayama, Shimane, Tottori, Yamaguchi, Ehime, Kagawa, Kochi, Tokushima, Fukuoka, Kagoshima, Kumamoto, Miyazaki, Nagasaki, Oita, Saga, Okinawa
Second-level Cluster Node	Child Nodes
Tohoku	Akita, Aomori, Fukushima, Iwate, Miyagi, Yamagata
KantoSub	Chiba, Gunma, Ibaraki, Kanagawa, Saitama, Tochigi
KansaiSub	Hyogo, Kyoto, Mie, Nara, Shiga, Wakayama
Leaf Nodes	All 47 prefectures

Table 4. Performance of SAHB-3LH-TWCNB on Prefectures

Training Classes	55 classes: 47 prefectures child nodes 8 region cluster nodes
Test Classes	47 prefectures
Correctly Classified Instances	10565 (94.558%)
Incorrectly Classified Instances	608 (5.442%)
Total Classified Instances	11173
Avg. Time to Classify one Instance	3,3 ms

5   Conclusion and Future Work

In this paper, we presented SAHB, an improvement to straightforward Naive
Bayes text classification methods, which is a meta-algorithm that can be trained
in an equivalent time to the subsequent classification algorithms it uses, and
takes advantage of a known hierarchy of classes as well as semantic proximity.

The idea for the algorithm stemmed from a broader project in which one of the research milestones is to classify by using the advantages of online open crowdsourced corpuses.

The algorithm needs further generalization and comparisons with other methods, and tests on other datasets, as it is limited in this case to a sample problem. Other cases include but are not limited to: having topic nodes belonging to multiple parents, bias such as equilibrium by clustering parent nodes together cannot be reached, which may require mixing with other techniques e.g oversampling. Finally, it also needs to be tested up to per-city classification in our example, although it is recursively intuitive that accuracy will still be better than straightforward use of TWCNB.

Acknowledgements. This work was supported and promoted by the Strategic Information and Communications R&D Promotion Programme (SCOPE)[1],[2], Ministry of Internal Affairs and Communications, Japan.

References

1. Dumais, S., Chen, H.: Hierarchical classification of web content. In: Proceedings of the 23rd Annual International ACM SIGIR Conference on Research and Development in Information Retrieval (SIGIR 2000), pp. 256–263. ACM, New York (2000)
2. Freund, Y., Schapire, R., Abe, N.: A short introduction to boosting. Journal-Japanese Society for Artificial Intelligence 14, 771–780 (1999)
3. Japkowicz, N., Stephen, S.: The class imbalance problem: A systematic study. Intell. Data Anal. 6(5), 429–449 (2002)
4. Rennie, J.D.M., Teevan, J., Karger, D.R.: Tackling the Poor Assumptions of Naive Bayes Text Classifiers. In: Proceedings of the Twentieth International Conference on Machine Learning, pp. 616–623 (2003)
5. Swezey, R., Shiramatsu, S., Ozono, T., Shintani, T.: Intelligent Page Recommender Agents: Real-Time Content Delivery for Articles and Pages Related to Similar Topics. In: Proceedings of the Twenty Fourth International Conference on Industrial, Engineering and Other Applications of Applied Intelligent Systems, IEA-AIE (2011)
6. Toutanova, K., Chen, F., Popat, K., Hofmann, T.: Text classification in a hierarchical mixture model for small training sets. In: Proceedings of the Tenth International Conference on Information and Knowledge Management (CIKM 2001), pp. 105–113. ACM, New York (2001)

---

[1] http://www.soumu.go.jp/main_sosiki/joho_tsusin /scope/index.html
[2] http://www.soumu.go.jp/soutsu/tokai/tool/koho siryo/hodo/22/ 08/0809-1.htm

Natural Language Processing

Intelligent Web-Based Natural Language Dialogue Planning for Web Information Filtering*

Hugo Hromic[1] and John Atkinson[2]

[1] DERI, National University of Ireland, Galway, Ireland
hugo.hromic@deri.org
[2] Department of Computer Sciences, Universidad de Concepcion
Concepcion, Chile
atkinson@inf.udec.cl

Abstract. This work proposes a stochastic web-based natural language dialogue planner for an intelligent web filtering model, which allows users to filter search results obtained by a traditional search engine and assists them to find what they really are looking for. Experiments with web users and different interaction settings show the promise of the approach to web-based adaptive planning.

1 Motivation

Current search engines have greatly improved the quality of the information retrieval task. However, the search process is still performed on blind basis respect to the *real interests* of the user and based solely on its ability to deliver keywords to a search engine.

In order to address the problem, a new approach for intelligent web documents search and *filtering* was proposed in the past few years [1]. It involved a natural language dialogue system to discover and refine the user's search *interests*. Thus, the dialogue system can greatly reduce the amount of results, by filtering those that are not relevant to the user. Through these natural language human-computer interactions, a system can interact with the user bidirectionally, confirming or rejecting assumptions that the system does or dynamically asking to refine the criteria and/or search concepts that the user needs.

Dialogue planning in this kind of systems becomes a key task to control and guide a dialogue. Its purpose is to plan the answers and interaction of the system with a user, by taking into account the current context of the dialogue, requests for information refinements, confirmations/rejections, specification of features to look for, etc. However, current dialogue planners take this kind of approach to intelligent web search filtering using rule-based methods, which are not efficient enough to deal with complex and dynamic dialogues with users, such as for web search and filtering.

---

* This research is sponsored by the educational technology project **FONDEF**, Chile under grant number **D08I1155**: *"Semantic Identification and Automatic Synthesis of Educational Materials for Specialised Domains"*.

W. Ding et al. (Eds.): Modern Advances in Intelligent Systems and Tools, SCI 431, pp. 155–160.
springerlink.com   © Springer-Verlag Berlin Heidelberg 2012

2 Dialogue Planning

Some approaches to dialogue planning propose a scheme to overcome the limitations of finite state-based approaches, including allowing mixed initiative dialog. For this, they use data structures similar to an electronic form (*E-Form*), which considers various constraints on the dialogue system's domain. To plan what the system should do, it processes the user input (a *frame* of information from a language understanding component) and then an algorithm based on a contextual E-Form of the dialogue is applied. This procedure provides facilities for information extraction from the input frame, combining this information with the contextual E-Form domain, data retrieval, contextual E-Form update and production of system responses. However, the model has still other limitations, such as flexibility in the interactions and the ability to adapt to unexpected responses.

To address these issues, a task-oriented planner for natural language dialogues has been proposed [5]. Here, asynchronous events that occur during the dialogue are specially treated. In order to handle dialogue asynchrony, the model uses an event queue for events which are received from other system modules. Instead of applying manually built rules or FSA, other approaches use statistical models to describe and predict sequences of dialogue moves. This kind of planner is trained with maximum likelihood estimators and a *corpus* of real labeled dialogues [3]. The model computes the next dialogue move that delivers the planner. Since there are many possible sequences of dialogues, a reduction of this search space is usually carried out. Thus, whenever an ongoing dialogue is not represented in the training corpus, a distance measure is used to determine the similarity of dialogue that is not seen within the training set.

Stochastic models have also been used successfully in areas of natural language processing (NLP) such as *Biomedical Entity Recognition, Information Extraction, Labeling Sequence of Words, Shallow parsing* and so on. On one side, *discriminative* Maximum Entropy Markov Models (MEMMs) exploit all the advantages of *generative* models. Although stochastic models have not been applied to dialogue planning, their ability to predict sequences make them good candidates to plan dialogue moves.

3 A CRF-Based Dialogue Planner

In this paper, a stochastic dialogue planning method based on Conditional Random Fields (CRF) is proposed. It was designed to be used into a dialogue system for Web document filtering. The CRF method allows great flexibility in guiding the dialogue, as it does not expect any defined particular response from the user and it is free to generate any answers to what users request at any time. In addition, this model is able to seize dialogue features that are not possible to capture with traditional generative or rule-based models.

The proposed planner takes *dialogue acts*[1] as input given by an associated natural language understanding component, and produces dialogue acts that will be the next dialogue move of the system, effectively working as a *dialogue act sequences predictor*.

The model's parameters and probabilities are estimated using a *training dialogue corpus*. These probabilities describe the candidate dialogue act sequences that might appear in Web search dialogue context. The planner also takes in account some dialogue *attributes* that are present anytime as the dialogue goes on, thus it uses this additional information for improving prediction. The current dialogue history is stored and updated everytime a new input is received. The output can be feed into a Natural Language Generation (NLG) module for creating direct user responses. In addition, the planner is also designed to interact with a Web search agent by building specific-purpose dialogue acts.

The model has also a single dialogue acts vocabulary for both input and output: SYS: for events issued by the *planner*, USR: for events issued by the *user* and SA: for events issued by the *search agent*.

In order to predict the next dialogue move, the CRF model *labels* dialogue acts in an input sequence (aka. observations), while computing the marginal probability of the complete label sequence. Thus, the planner takes every dialogue act in the model vocabulary (the *candidate* acts), appends one by one temporarily to the current dialogue history and the entire candidate sequence is labeled using the CRF model. Finally, the sequence with the highest probability is chosen and the related candidate dialogue act is selected as the next dialogue move for output.

The model has a single dialogue acts vocabulary for both input and output (such as GREET, REQ_TOPIC, ANSWER_TOPIC, etc.), so it needs to prefix those acts with the originating entity with the following strings: SYS : for events issued by the planner, USR : for events issued by the user and SA : for events issued by the search agent. As previously mentioned, the planner also keeps the history sequence of dialogue (history). The next candidate dialogue act is SYS:SYS_SEARCH and the CRF planner will compute the marginal probability of the entire sequence. If this probability is greater than all other candidate dialogue acts on the model vocabulary, it will be chosen as the output dialogue move.

CRF models on *dialogue attributes* are used to increase available information to make next dialogue move predictions. Dealing with dialogue attributes is realized into two steps: a) labeling the training corpus, and b) labeling on running a dialogue session. Different features are defined to model the dialogue on stochastic planner: **Topic Set (TS)** (indicates whether a particular point of the dialogue the user has given a search topic), **System Searching (SS)** (indicates whether the search agent is currently searching the Web), **Results Found (RF)** (indicates whether results have been obtained by the search agent and also they are in the system's memory), **Dialogue Phase (PHASE)** (indicates the current stage of the dialogue being held at a given time).

---

[1] A kind of *action* associated to someone's expression, regardless of how it's uttered.

In the case of the training set, attributes are part of the manual labeling of the corpus, whereas for normal operation, *rules* were designed to assign values to different attributes during the course of the dialogue. These rules are activated whenever a dialogue movement occurs. For example, the dialogue act: USR:ANSWER_TOPIC produces the attribute updates: TS ← YES, PHASE ← WAIT_RESULTS, and leaving the rest unchanged.

Dialogue attributes on our stochastic planner are incorporated by using *feature functions*. *Attribute templates* are defined and automatically expanded to generate different feature functions of the dialogue to train the parameters of the model, given a set of attribute-value pairs.

4   Results

Our approach to stochastic dialogue planning was evaluated for performance against other traditional dialogue planning models such as a *n-gram* model and a *finite-state* based planner (the baseline). The CRF model can be tuned by the regularization method, hyper-parameter values [2], the minimum frequency of attributes to consider during training, and dialogue attribute sets to take in account for planning.

We also measured the number of interactions (dialogue turns) taken by the user, the planner and the search agent until the task is completed, the quality of dialogue adjacency pairs and the dialogue act diversity that each planner is capable of delivering.

For training the statistical models (CRF and n-gram), an annotated dialogue corpus was created using the *Wizard of Oz* methodology [4]. This included 64 transcriptions of different interactions of users with a dialogue system to search for information on the Web.

Overall, 10 volunteer users ranging in age from 20 to 50 years old were considered for the WoZ sessions. All dialogues involved assisting and filtering information for users looking for common topics on the web, in which the main search themes included: *leisure, science and technology, politics and news, health and psychology*, and *culture, cinema and books*.

CRF models were generated using the *CRF++*[2] tool which implements iterative training based on L-BFGS optimization [6], type L1 and L2 regularization, feature function generation using templates and is able to compute the marginal probabilities of the candidate dialogue sequences. In order to make the assessment a *k-fold* ($k = 5$) type cross-validation was used.

In order to compare planners, n-gram and finite-state planners were used. An extrinsic evaluation metric was designed to score the *overall combined performance* of the planners, called *Dialogue Planner Rank (DPR)*. DPR is a linear combination that considers various issues of dialogue which aimed to assess the communicative and dialogue efficiency: the number of turns used in a dialogue session, the number of times required to access the search agent, the dialogue acts diversity and the ability to correctly predict the next dialogue move.

---

[2] http://crfpp.sourceforge.net/

The proposed DPR can be seen in equation 1, where ady_num is the number of dialogue adjacency pairs that appears during the session, ady is the set of these pairs, ady_pos_i is the position of the adjacency pair i in the sub-list of expected dialogue acts (which closer to 1 is better), $turns$ the number of dialogue turns, $search$ the number of performed Web searches, $acts\_shown$ the number of different dialogue acts that the planner showed and $total\_acts$ the total number of dialogue acts in the vocabulary of the planner.

$$DPR = (\frac{1}{ady\_num} * \sum_{i \in ady} \frac{1}{ady\_pos_i}) + \frac{1}{turns} + \frac{1}{searches} + \frac{acts\_shown}{total\_acts} \quad (1)$$

Computing the expected dialogue acts score for each adjacency pair was carried out from 30 dialogues randomly chosen from the training corpus and 20 additional dialogues generated by the WOZ methodology. Thus, the three planners (CRF, n-gram and FSA) were fed with these dialogues, calculating their corresponding DPR and obtaining the results shown in figure 1.

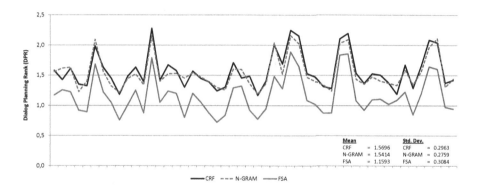

Fig. 1. DPR for different evaluated web-driven dialogue planners

The figure shows that statistical planners (CRF and n-grams) clearly outperform the traditional rule-based planner (FSA), getting better DPR in all dialogues. At the same time, statistical planners show a performance very similar to each other, with CRF being a bit better than n-gram planners. In fact, the DPR averages shows that our CRF planner surpasses the n-gram model.

In general, the planner does not need to increase access to the search engine (a significant reduction of processing cost). The proposed CRF dialogue planning model also generates a greater diversity of dialogue acts responses than the n-gram model, which results in more interesting and spontaneous dialogue with the system, as the CRF planner is capable of capturing, through the dialogue attributes, the different dialogue points within a conversation with the user. Overall, the results showed that a CRF-based dialogue planner can be effective to allow web search and filtering dialogue managing with users looking for information on the web.

5   Conclusions

In this paper, a new stochastic planner was proposed to assist web users when searching for information on the web, by filtering contents using natural language dialogues. The proposed model was evaluated against traditional planning techniques (n-gram models and FSA) so as to measure efficiency and effectiveness to complete the dialogue tasks.

Assessing the model under different metrics included the number of interactions, access times to the web search engine, planned dialogue acts diversity and quality of planned dialogue adjacency pairs. Experiments showed that CRF-based planners and statistical (n-gram) planners do not require a detailed study of the dialogues in the corpus, since they can automatically learn from it.

Our approach outperformed other planners even requiring no defined schemata to interact with the user. This is mainly due to the discriminative nature of the CRF model which allows it to get rid of strong dependency assumptions in the dialogue sequences, and to incorporate attributes that other generative models can not, effectively increasing information available to make predictions.

References

1. Ferreira, A., Atkinson, J.: Intelligent search agents using web-driven natural-language explanatory dialogs. IEEE Computer, 44–52 (October 2005)
2. Friedman, J., Hastie, T., Tibshirani, R.: Regularization paths for generalized linear models via coordinate descent. Department of Statistics. Stanford University (2008) (manuscript)
3. Hurtado, L.F., Griol, D., Sanchis, E., Segarra, E.: A stochastic approach to dialog management. In: IEEE Workshop on Automatic Speech Recognition and Understanding, pp. 226–231 (October 2005)
4. Jurafsky, D., Martin, J.H.: Speech And Language Processing: An Introduction to Natural Language Processing, Computational Linguistics and Speech Recognition. Prentice-Hall (2008)
5. Stallard, D.: Dialogue management in the Talk'n'Travel system. In: IEEE Workshop Automatic Speech Recognition and Understanding (ASRU 2001), pp. 235–239 (2001)
6. Wallach, H.M.: Conditional Random Fields: An introduction. Technical Report MS-CIS-04-21. Department of Computer and Information Science, University of Pennsylvania, 50 (2004)

Enhancing Text Classifiers to Identify Disease Aspect Information

Rey-Long Liu

Department of Medical Informatics
Tzu Chi University, Hualien, Taiwan
rlliutcu@mail.tcu.edu.tw

Abstract. Etiology, diagnosis, treatment, prevention, and symptoms are key aspects of disease information. Automatic identification of medical texts about the aspects is essential since (1) disease aspect information is routinely created and updated by many different sources, and (2) both healthcare professionals and consumers need to retrieve the disease aspect information. Main challenges of identifying the disease aspect information include (1) many medical texts are written for a specific disease and hence often contain information about multiple aspects of the disease, and (2) disease aspects are predefined categories of interest, making it difficult to precisely express the interest with brief descriptions. In this paper, we model the aspect identification problem as a text classification problem, and develop a technique IDAI (Identification of Disease Aspect Information) that considers term contexts to enhance text classifiers to classify medical texts with information about multiple disease aspects (categories). Empirical results show that IDAI enhances a Support Vector Machine (SVM) classifier, and when compared with an enhancement technique that considers term contexts, IDAI performs better. The contribution is of significance to text classification technology, evidence-based medicine, and health education and decision support.

1 Introduction

Healthcare information for specific diseases is essential for both healthcare professionals and consumers. Key *aspects* of the information include *etiology*, *diagnosis*, *treatment*, *prevention*, and *symptoms* of specific diseases, which are often required in clinical practice, health education, and health promotion (Liu, 2007; Lin & Demner-Fushman, 2006). Therefore, healthcare information providers (e.g., MedlinePlus[1]and MedicineNet[2]) often need to routinely spend much effort to collect, compile, and update the disease aspect information for archiving and retrieval.

In this paper, we develop a technique IDAI (Identification of Disease Aspect Information) to enhance various kinds of text classifiers to classify the medical text into

---

[1] Available at http://www.nlm.nih.gov/medlineplus/healthtopics.html
[2] Available at http://www.medicinenet.com

W. Ding et al. (Eds.): Modern Advances in Intelligent Systems and Tools, SCI 431, pp. 161–166.
springerlink.com   © Springer-Verlag Berlin Heidelberg 2012

multiple aspects (categories). The development of IDAI is motivated by two common phenomena in practice: (1) disease aspect information is routinely created and updated by many different sources, and (2) both healthcare professionals and consumers need to retrieve the disease aspect information (e.g., for a disease, healthcare professionals search for diagnosis and treatment evidences, while healthcare consumers search for symptoms and prevention guidelines).

Main challenges of identifying disease aspect information include (1) disease aspects are predefined categories of interest, making it difficult to precisely express the interest with brief descriptions, and (2) many medical texts are written for a specific disease and hence often contain information about multiple aspects of the disease. For the former, IDAI models the aspect identification problem as a text classification problem, and hence no precise descriptions for each aspect are required. For the latter, IDAI enhances classifiers by helping them to consider the *context* of terms in the text, based on the observation that the terms related to an aspect often *collectively* appear in some segments of the text.

2   Related Work

To our knowledge, IDAI is the first attempt to modeling the problem of disease aspect identification as a text classification problem. Previous text classification studies have developed many classifiers among which the Support Vector Machine (SVM) classifier have been routinely employed (e.g., Bennett & Nguyen, 2009) and often shown to be one of the best classifiers. The category of a text is determined by classifying the whole text, or by merging the categories of individual parts of the text (Kim & Kim, 2004). However, the aspect classification problem introduces an impact to the classifiers, since a medical text may have several parts about different disease aspects. If a text *simultaneously* mentions features related to *different* categories, features related to other categories may lead the classifiers *not* to classify the text into c, even though the text has features for c.

One straightforward method to categorize textual parts of a text is to split the text into several parts and then classify the parts individually. The method was used to detect those parts that are semantically quite different from the whole text (e.g., detecting the "hidden" passages in the texts about business, Mengle & Goharian, 2009). However, the method is not suitable for disease aspect classification either, since disease aspects are semantically correlated, making it quite difficult to define and tune a proper way to set the location and the length of each part for each aspect. Splitting a text into several parts for the disease aspects may thus even be another aspect classification problem.

Previous techniques to find textual parts from a text include topic segmentation, topic summarization, topic tracking, and passage retrieval. However, they focused on detecting the existence and/or transition of topics in texts, as well as retrieving those textual parts that are related to a given topic description or question. They did not consider disease aspects, which are predefined *categories* of interest, and hence cannot be precisely expressed with brief descriptions or questions.

3   IDAI

IDAI employs term context information to revise term frequency (TF) of each term t in a text d. The TF of each term is then input to the underlying classifier to enhance the performance of the classifier in aspect classification. As TF is commonly considered by many classifiers, IDAI may collaborate with the classifiers.

In training, IDAI employs χ^2 (chi-square) statistics to identify the type of correlation between each term and category. For a term t and a category c, $\chi^2(t,c) = [N \times (A \times D - B \times C)^2] / [(A+B) \times (A+C) \times (B+D) \times (C+D)]$, where N is the total number of training texts, A is the number of training texts that are in c and contain t, B is the number of training texts that are not in c but contain t, C is the number of training texts that are in c but do not contain t, and D is the number of training texts that are not in c and do not contain t. A term t is *positively correlated* to a category c, if $A \times D > B \times C$; otherwise it is *negatively correlated* to c.

In testing, IDAI employs the term-category correlation types to revise the TF value of each term with respect to each category. To implement the TF revision, we extend the context-based technique developed by Liu, 2010, and Fig. 1 defines the equations used by IDAI to compute the revised TF (RTF) of each feature term. For a term t that is *positively* correlated to a category c, its TF is estimated by considering the term windows centered by each occurrence of t in d (*WindowTF* in Equation 1). For each occurrence of t at position k (i.e., t_k), IDAI adds 0.5 and a window proximity score $P_{window}(t_k,d,c)$ to *WindowTF* of t (see Equation 2). The window proximity score is in the range of [0, 1], and hence if the score < 0.5, the TF is actually decreased (since in traditional TF assessment, each occurrence of t increases the TF of t by 1.0). $P_{window}(t_k,d,c)$ indicates the extent to which other terms that are positively correlated to c co-occur with t_k in the term window. It is measured by *WinContext* and *MaxWinContext* values of t_k in d with respect to c (see Equation 3). *WinContext*(t_k,d,c) is the sum of the proximity strengths of neighboring terms of t_k that are positively correlated to c and fall in a term window around k (see Equation 4). *MaxWinContext* is defined to normalize $P_{window}(t_k,d,c)$ into the range of [0,1] (see Equation 5).

On the other hand, for a term t that is *negatively* correlated to a category c, its TF is estimated by considering two factors: (1) the maximum *WindowTF* with respect to some category other than c, and (2) an inconsistent correlation type score (see *InconsistencyTF* in Equation 1). The inconsistent correlation type score is defined in Equation 6 to Equation 8. IDAI reduces the TF of a term t that is negatively correlated to c if t occurs after a text segment in which those terms that are positively correlated to c dominate (see Equation 7 and Equation 8).

[1] $RTF(t,d,c) =$

   $WindowTF(t,d,c)$, if t is positively correlated to c;

   $Max_{c_i \neq c}\{WindowTF(t,d,c_i)\} - InconsistencyTF(t,d,c)$, otherwise.

[2] $WindowTF(t,d,c) =$

   $\sum_{t_k} 0.5 + P_{window}(t_k,d,c),$ where t_k is an occurrence of t at position k in d

[3] $P_{window}(t_k, d, c) =$

   0, If none of the immediate neighbors of t_k in d is positively correlated to c;

   $WinContext(t_k,d,c) / MaxWinContext(t_k,d,c)$, otherwise.

[4] $WinContext(t_k,d,c) = \sum_{n_k} \dfrac{0.9 * dist(n_k,t_k) - winsize + 0.1}{1 - winsize}$, where n_k is a neighboring term

whose distance to t_k (i.e., $dist(n_k,t_k)$) is not large than the window size $winsize$ (set to 5), and n_k is positively correlated to c.

[5] $MaxWinContext(t_k, d, c) =$ Maximum possible $WinContext(t_k,d,c)$ for t_k.

[6] $InconsistencyTF(t,d,c) = \sum_{t_k} P_{inconsist}(t_k,d,c)$

[7] $P_{inconsist}(t_k, d, c) =$

   0, if $NetPosNum(k,d,c) \leq MinPosNum$;

   0.5, if $NetPosNum(k,d,c) \geq MaxPosNum$;

   $0.5 \times \dfrac{NetPosNum(k,d,c) - MinPosNum}{MaxPosNum - MinPosNum}$, otherwise, where $MinPosNum$ (set to 3)

   and $MaxPosNum$ (set to 20) are two parameters governing the minimum and maximum $NetPosNum$ values, respectively.

[8] $NetPosNum(k,d,c) = Max_{1 \leq b \leq k}\{\sum_{m=k}^{b} CtypeS(m,c)\}$, where $CtypeS(m,c) = 1$ if the feature

appearing at position m in d is *positively* correlated to c; otherwise $CtypeS(m,c) = -1$.

Fig. 1. Revised term frequency (RTF) estimated by IDAI

4   A Case Study on Chinese Texts of Disease Aspects

To evaluate IDAI, we conducted a case study on thousands of Chinese texts describing cancers and top fatal diseases in Taiwan. In the case study, top-10 fatal diseases and top-20 cancers in Taiwan were considered, resulting in 28 diseases of interest. Chinese texts about the 28 diseases were collected from various sources, including the web sites of National Taiwan University Hospital[3], Department of Health in Taiwan[4], and many related associations and hospitals. For each disease, we were interested in 5 aspects (categories): *etiology, diagnosis, treatment, prevention,* and *symptom.* As noted in Section 1, a medical text about a disease often contains information about several aspects for the disease. Therefore, we carefully checked each medical text so that the text could be split into several texts about disease aspects, resulting 4669 texts.

---

[3] Available at http://www.ntuh.gov.tw/en/default.aspx
[4] Available at http://www.doh.gov.tw/EN2006/index_EN.aspx

We randomly sampled 10% of the 4669 texts for testing (classification), and the other texts for training. As noted above, we have confirmed that in practice many medical texts mention information of multiple aspects. Therefore, to experiment on the practical environment, for each test document we randomly combined it with other test documents that were about different aspects of the same disease, resulting in 468, 467, 467, 467, and 429 documents that mentioned 1, 2, 3, 4, and 5 aspects, respectively.

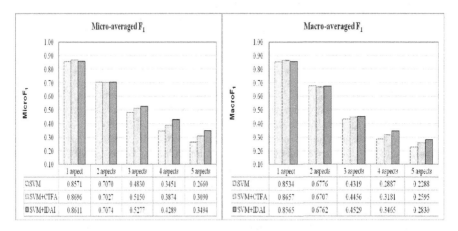

Micro-averaged F_1	1 aspect	2 aspects	3 aspects	4 aspects	5 aspects
SVM	0.8571	0.7070	0.4830	0.3451	0.2660
SVM+CTFA	0.8696	0.7027	0.5150	0.3874	0.3090
SVM+IDAI	0.8611	0.7074	0.5277	0.4289	0.3494

Macro-averaged F_1	1 aspect	2 aspects	3 aspects	4 aspects	5 aspects
SVM	0.8534	0.6776	0.4319	0.2887	0.2288
SVM+CTFA	0.8657	0.6707	0.4456	0.3181	0.2595
SVM+IDAI	0.8565	0.6762	0.4529	0.3465	0.2830

Fig. 2. IDAI enhances SVM, and performs better than CTFA in enhancing SVM to classify those texts that contain multiple aspect information

To measure the performance of aspect identification (text classification), we employed two criteria: *micro-averaged F_1* (MicroF$_1$) and *macro-averaged F_1* (MacroF$_1$), which are routinely employed in previous studies (Liu, 2007). Moreover, to build an underlying classifier to be enhanced, SVM[Light] was employed to build an SVM classifier (Joachims, 1999). To further measure the contribution of IDAI with respect to related techniques, we also implemented a baseline enhancer CTFA (Liu, 2010) as IDAI is extended from CTFA (as noted in Section 3), which employs term context information to enhance classifiers as well.

The results indicate that performance of aspect classification becomes much poorer when the texts mention several disease aspects (see Fig. 2). IDAI successfully enhances SVM to classify those texts that contain multiple aspect information Moreover, IDAI performs better than CTFA in enhancing SVM to classify those texts that contain multiple aspect information as well, justifying the contribution the second kind of TF revision conducted by IDAI.

5   Conclusion and Future Work

Automatic identification of disease aspect information is essential for disease information archiving, retrieval, and comparison to facilitate evidence-based medicine, health education, and healthcare decision support. In the case study on cancers and top fatal diseases in Taiwan, the technique IDAI proposed in the paper performs better than a technique in enhancing a state-of-the-art classifier to identify the medical texts about the disease aspects. It is interesting to extend IDAI to further locate, in the medical texts being classified, the positions of the textual parts about each disease aspect of interest.

Acknowledgments. This research was supported by the National Science Council of the Republic of China under the grant NSC 100-2221-E-320-004-MY2.

References

1. Bennett, P.N., Nguyen, N.: Refined Experts Improving Classification in Large Taxonomies. In: Proceedings of the 32nd Annual International ACM SIGIR Conference on Research and Development in Information Retrieval, Boston, USA, pp. 11–18 (2009)
2. Joachims, T.: Making Large-Scale SVM Learning Practical. In: Schölkopf, B., Burges, C., Smola, A. (eds.) Advances in Kernel Methods - Support Vector Learning. MIT-Press (1999)
3. Kim, J., Kim, M.H.: An Evaluation of Passage-Based Text Categorization. Journal of Intelligent Information Systems 23(1), 47–65 (2004)
4. Lin, J., Demner-Fushman, D.: The role of knowledge in conceptual retrieval: a study in the domain of clinical medicine. In: Proceedings of the 29th Annual International ACM SIGIR Conference on Research and Development in Information Retrieval, Seattle, Washington, USA, pp. 99–106 (2006)
5. Liu, R.-L.: Context-based Term Frequency Assessment for Text Classification. Journal of the American Society for Information Science and Technology 61(2), 300–309 (2010)
6. Liu, R.-L.: Text Classification for Healthcare Information Support. In: Okuno, H.G., Ali, M. (eds.) IEA/AIE 2007. LNCS (LNAI), vol. 4570, pp. 44–53. Springer, Heidelberg (2007)
7. Mengle, S., Goharian, N.: Passage Detection Using Text Classification. Journal of the American Society for Information Science and Technology 60(4), 814–825 (2009)

Text Clustering on Oral Conversation Corpus

Ding Liu and Minghu Jiang

Lab. of Computational Linguistics, School of Humanities and Social Sciences,
Tsinghua University, Beijing 100084, China
shuidiaoxiang@163.com, jiang.mh@tsinghua.edu.cn

Abstract. This article describes a method that use some context information terms in text clustering base on oral conversation corpus. And we used various distance measurement in the SOM algorithm experiment and the K-means algorithm experiment to test it. The experimental results showed us the context information terms take effect on text clustering, because of its high occurrence frequency. And we found that Hamming distance measurement is the best choice in SOM algorithm.

Keywords: text clustering, context information terms, oral conversation corpus, SOM, K-means, distance measurement.

1 Introduction

The study on corpus brings some new ideas and new methods for computational linguistics and traditional linguistics in recent years. Researchers use lots of news corpora, literature corpora, but the oral conversation corpus base on the daily conversation is rarely used in study. This kind of corpus consists of daily conversation which is rich in the participator information and context information. And this kind of information could be considered as a part of pragmatics information, which is rarely used in recent study. Therefore, how to utilize pragmatics information in nature language processing will be a new field that worth to be concerned.

2 Context and Oral Conversation Corpus

"Context" is an important concept in pragmatics system, which is proposed by linguist Malinowski in 1923. After that, linguist Firth accepted Malinowski's thinking and constructed the whole theory of context. He explained the context as follows in his book *A Synopsis of Linguistic Theory.*[12]

We find that the "participator" and the "situation" are two important concepts in context theory, and the feature of common conversation text is decided by them to some extent. But the "situation" is too abstract, and difficult to be formulized for computation, therefore we replace it by the concept "scene". Oral conversation text

W. Ding et al. (Eds.): Modern Advances in Intelligent Systems and Tools, SCI 431, pp. 167–172.
springerlink.com

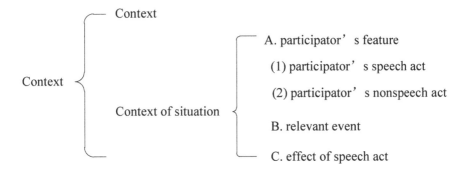

Fig. 1. Context Theory

contains more participator information and scene information than news and literature text.Therefore we used the oral conversation corpus to extract participator information and scene information in our experiment, and used them in text clustering.

3   Text Clustering and Distance Measurement in Clustering Algorithm

The study on text clustering has a long history, and applied widely in information retrieval, search engine, text mining and so on. Researchers have developed lots of clustering algorithm, and we adopted the Self-Organization Map algorithm(SOM) and K-means algorithm in our experiment, both of them are unsupervised learning algorithm that researcher often used.

Generally, the SOM algorithm use Euclidean distance measurement to compute the distance between output neuron nodes and input vectors, and K-means algorithm also use it to compute the similarity between vectors. But the Euclidean distance measurement doesn't lead to a good result in oral conversation text clustering experiment, whatever in SOM or K-means algorithm. Therefore, we compared some distance measurements which are often used in data analyses in our experiment, and the results will be explained later. We introduce distance measurements which are listed as follows[13].

Table 1. Distance measurements

Euclidean	$d_{st}^2 = (x_s - x_t)(x_s - x_t)'$
SEuclidean	$d_{st}^2 = (x_s - x_t)V^{-1}(x_s - x_t)'$
CityBlock	$d_{st} = \sum_{j=1}^{n} \left| x_{sj} - x_{tj} \right|$

Table 1. *(continued)*

Minkowski	$d_{st} = \sqrt[p]{\sum_{j=1}^{n} \left| x_{sj} - x_{tj} \right|^{p}}$
Chebychev	$d_{st} = \max_{j} \left\{ \left| x_{sj} - x_{tj} \right| \right\}$
Correlation	$d_{st} = 1 - \dfrac{\left(x_{s} - \bar{x}_{s} \right)\left(x_{t} - \bar{x}_{t} \right)'}{\sqrt{\left(x_{s} - \bar{x}_{s} \right)\left(x_{s} - \bar{x}_{s} \right)'}\sqrt{\left(x_{t} - \bar{x}_{t} \right)\left(x_{t} - \bar{x}_{t} \right)'}}$
Hamming	$d_{st} = (\#(x_{sj} \neq x_{tj}) / n)$
Cosine	$d_{st} = 1 - \dfrac{x_{s} x_{t}'}{\sqrt{\left(x_{s} x_{s}' \right)\left(x_{t} x_{t}' \right)}}$

4   Experimental Analyses

We adopted the oral conversation corpus which collected by the computational linguistics lab of Tsinghua university. The corpus consists of 5 kinds of texts, each of them contains 40 texts, amount to 200 texts .We show it as follows.

Table 2. The oral conversation corpus

mark	A	B	C	D	E
content/scene	See doctor	Have dinner	Buy ticket	By book	At bank
quantity	40	40	40	40	40

Experimental flow chart is shown as follows:

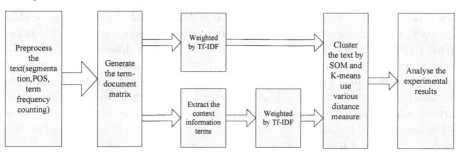

Fig. 2. Experimental flow chart

Context information term means the words identify participator and scene as we discussed before. After preprocessing and clearing the text, the term-document matrix contains 1371 terms, and the number of context information terms we selected is 88. The occurrence frequency of these context information terms is higher than others, therefor we could pick up them by selected the high occurrence frequency words, and then checked them by hand.

We used 5*1 map structure in SOM experiment, and clustered the sample into 5 groups in K-means experiment. It is convenient to calculate the F-measure and analyses experimental results. We used the F-measure method to evaluate the experimental results, the principle is shown as follows[11]:

$$\Pr ecision(P_j, C_i) = \frac{|P_j \cap C_i|}{|C_i|} \tag{1}$$

$$\mathrm{Re}\, call(P_j, C_i) = \frac{|P_j \cap C_i|}{|P_j|} \tag{2}$$

$$F(P_j, C_i) = \frac{2P(P_j, C_i) * R(P_j, C_i)}{P(P_j, C_i) + R(P_j, C_i)} \tag{3}$$

$$F(P_j) = \max_{i=1,2\dots m} \{F(P_j, C_i)\} \tag{4}$$

$$F = \frac{\sum_{=1}^{s}(|P_j| * F(P_j))}{\sum_{j=1}^{s}|P_j|} \tag{5}$$

In SOM experiment, we trained the net 300 times, and then calculated the F-measure. In the K-means experiment, we tested 10 times for each distance measurement, because k-means algorithm is an unstable learning algorithm, then we calculated the average F-measure. The results are shown as follows.

According to SOM experimental results, we find that Hamming distance lead to the best results, the SEuclidean distance, Correlation distance and Cosine distance lead to the worst when we use the whole term-document matrix. And the context information terms improve results apparently when we use Euclidean distance, but it lead to F-measure reduction when we use Hamming distance, because of the appearance of zero vectors in some samples. In the K-means experiment, we find that Correlation distance and Cosine distance lead to better results when we use the whole term-document matrix, but both of them lead to failure when we use the context information terms, because of the appearance of zero vectors in some samples. But context information terms improve results to some extent when we use Euclidean distance and Cityblock distance.

Table 3. The SOM experimental results (300 iterations)

	Use whole terms	Use context information terms
Euclidean	0.5552	0.8484
SEuclidean	0.3333	0.3333
CityBlock	0.7317	0.7266
Minkowski	0.7686	0.7041
Chebychev	0.7569	0.7486
Correlation	0.3333	0.3333
Hamming	0.9899	0.8701
Cosine	0.3333	0.3333

Table 4. The K-means experimental results (average F-measure by 10 times)

	Use whole terms	Use context information terms
Euclidean	0.4994	0.6358
CityBlock	0.4463	0.5289
Correlation	0.9193	/
Cosine	0.9354	/

The histogram is shown as follows.

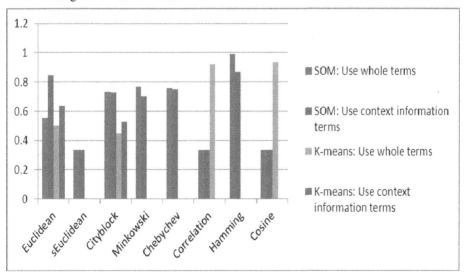

Fig. 3. The histogram of experimental results

5   Conclusion

Context information term leads to various results depend on different clustering algorithm and different distance measurement. One of the reasons is different distance measurement has different sensitivity on high occurrence frequency term and zero vector. And the task that utilize date mining or other machine learning method to extract more pragmatics information from corpus, then use it to help text clustering is worth researching in the next step.

Acknowledgement. This work was supported by the National Natural Science Foundation in China (61171114), State Key Lab of Pattern Recognition open foundation, CAS. Tsinghua University Self-determination Research Project (20111081023&20111081010) and Human & liberal arts development foundation (2010WKHQ009).

References

1. Ampazis, N., Perantonis, S.J.: LSISOM – A Latent Semantic Indexing Approach to Self-Organizing Maps of Document Collections. Neural Processing Letters 19, 157–173 (2004)
2. Bolasco, S., Canzonetti, A., Capo, F.M., et al.: Understanding Text Mining: A Pragmatic Approach. In: Sirmakessis, S. (ed.) Knowledge Mining. StudFuzz, vol. 185, pp. 31–50. Springer, Heidelberg (2005)
3. Honkela, T.: Self Organizing Map. In: Natural Language Processing. PhD thesis. Helsinki University of Technology, Neural Networks Research Centre, Helsinki
4. Kohonen, T., Kaski, S., Lagus, K., et al.: Self Organization of a Massive Document Collection. IEEE Transactions on Neural Networks 11(3) (May 2000)
5. Kohonen, T., Makisara, K.: The Self-organizing Feature Maps. Physica Scripta 39, 168–172 (1989)
6. Lagus, K., Honkela, T., Kaski, S.: WEBSOM for Textual Data Mining. Artificial Intelligence Review 13, 345–364 (1999)
7. Lagus, K., Kaski, S., Kohonen, T.: Mining massive document collections by the WEBSOM method. Information Sciences 163, 135–156 (2004)
8. Merkl, D.: Text classiÞcation with self-organizing maps:Some lessons learned. Neuro Computing 21, 61–77 (1998)
9. Ritter, H., Kohonen, T.: Self-Organizing Semantic Maps. Biological Cybernetics 61, 241–254 (1989)
10. Zeng, W.: Multi-perspectives on Pragmatics. ZheJiang University Press, Hangzhou (2009)
11. Zhou, Z.: Quality Evaluation of Text Clustering Results and Investigation on Text Representation. M.E.thesis. Institute of Computing Technology Chinese Academy of Sciences (2005)
12. 索振羽.语用学教程.北京.北京大学出版社 (2000)
13. Function Documentation of Matlab,
    http://www.mathworks.cn/help/toolbox/stats/pdist.html

A Semantic Relevancy Measure Algorithm of Chinese Sentences

Yan Chen, Yang Yang, and Haiping Zhu

Department of Computer Science and Technology
Xi'an Jiaotong University, 710049, China
MOE KLINNS Lab and SPKLSTN Lab
Xi'an Jiaotong University, 710049, China
{chenyan,zhuhaiping}@mail.xjtu.edu.cn, sunny51@foxmail.com

Abstract. The semantic relevancy measures between sentences play an increasingly important role in text-related research and applications in areas such as text categorization, text-reasoning, text structure analysis and Question-Answering system. In this paper, focusing on the Chinese short text, a novel semantic relevancy measure algorithm between sentences is proposed. This method calculates sentence relevancy by combining the word-form feature, semantic feature and syntax feature in sentences. Besides semantic feature, the syntax structure information of sentences is also considered. Experiments prove that the proposed algorithm is efficient and useful in semantic relevancy measure of Chinese sentence.

Keywords: Semantic Relevancy, Syntax Feature, Semantic Dependency Grammar, Multi-feature Combination.

1   Introduction

Recently, applications of natural language processing present a need for an effective method to compute the relevancy between short texts or sentences. For example, sentence relevancy plays a very important role in the topic detection and tracking system for interactive text [1]. The topic correlation detection method is related to sentence relevancy. In text mining, sentence relevancy is used as a criterion to discover unseen knowledge from textual databases. In addition, the incorporation of short-text relevancy is beneficial to applications such as text summarization [2], text categorization [3], and Question-Answering system [4]. These exemplar applications show that the computing of sentence relevancy has become a generic component for the research community involved in text-related knowledge representation and discovery.

Although there are few publications relating to the measurement of relevancy between very short texts or sentences, as described in the related work section in this paper, the adaptation of available measures to computing sentence relevancy has its drawbacks. All of them are focusing on the semantic of words in a sentence, but missing the syntax structure information of sentences. Aiming at avoiding this drawback, we propose a new approach based on multi-features to measure semantic relevancy of Chinese sentence.

W. Ding et al. (Eds.): Modern Advances in Intelligent Systems and Tools, SCI 431, pp. 173–178.
springerlink.com   © Springer-Verlag Berlin Heidelberg 2012

The rest of this paper is organized as follows: First, Section 2 presents related works among sentence relevancy researches. In the following, Section 3 introduces what sentence relevancy is and the details about the semantic relevancy measure algorithm. Section 4 describes the process of experiments and analyzes the results. At last, we discuss the future work in Sections 5.

2   Related Work

This section reviews several related work in order to explore the strengths and limitations of previous methods, and to identify the particular difficulties in computing sentence relevancy. Jagadeesh used relevance based language modeling to scores the sentences and then present a sentence extraction based summarization system [2]. To get the optimal answer for the query in the Question-Answering system, Li Sujian presented an approach to calculate the relevancy between two sentences based on Cilin and hownet [4]. ZHANG Youhua proposed a text category model based on sentence correlation which was obtained by means of sentence position weight and corpus item weight to achieve correlation matrix for text classification [3]. ZHONG Maosheng [5] divided sentence relevancy measure methods into two ways, qualitatively and quantitatively.

Therefore, we can make a conclusion that the most essential part in sentence relevancy research is correlation between words. Related works can roughly be classified into two major categories: knowledge-based methods [6] , and corpusbased methods, which are based on statistic analysis of large scaled corpus [7] . From a general point of view, knowledge-based methods are more simple and intuitive, but the results are always subjective. Methods based on large scaled corpus are objective while the sparse and noise data could result in unavoidable interferences. In this paper, we proposed a method both on knowledge dictionary and large corpus so that we can get an objective and wide word pair coverage result.

3   Sentence Relevancy Calculation

Sentence relevancy is mainly about whether the two sentences are about the same event. In our paper, a multi-features combination method is used to calculate the sentence correlation. While the multi-features are the keywordform feature, the semantic feature and the syntactic feature. Both of the word semantic information and the sentence structure information are considered. The details of the method are showed as following.

As for two sentences A and B, we suppose that A_1, A_2, …, A_M are the keywords in A and B_1, B_2, …, B_N are the keywords of B.

The relevancy upon the keyword-form feature is formRele(A,B).

$$\text{formRele}(A, B) = 2 * SameWord(A, B) / (Len(A) + Len(B))$$

formRele(A,B) is the number of the words that both in A and B. Len(A) and Len(B) are the number of the words belonged to A and B.

TongYiCi Cilin [8] is imported in the proposed method of semantic correlation. The correlation between two words is assumed as the nearest distance in the semantic tree of TongYiCi Cilin. The keyword correlation matrix between two sentences is built upon this method. Then, we use large scaled corpus to revise the matrix.

semanticRele(A,B) is the semantic relevancy upon the keyword semantic feature.

$$semantic \operatorname{Re} le(A,B) = (\frac{\sum\limits_{i=1}^{m} a_i}{m} + \frac{\sum\limits_{i=1}^{n} b_i}{n})/2$$

$$a_i = \max(s(A_i, B_1), s(A_i, B_2), ..., s(A_i, B_n)) \quad b_i = \max(s(B_i, A_1), s(B_i, A_2), ..., s(B_i, A_m))$$

$s(A_i, B_j)$ is the word relevancy in semantic feature between word A_i and B_j.

Semantic dependency grammar is used to calculate the correlation on the syntactic feature [9]. The effective collocation pairs of the dependent structure are considered in the correlation calculation. An effective collocation pair refers to the core word and all the effective words directly depending on it in a sentence while the effective words here must be verbs, nouns and adjectives after participle and part-of-speech tagging.

To combine the multi-features, we sum the three relevancies with different weights together. However, different weights are given to different features according to their contributions to the sentence relevancy. In order to find out the weight of each feature accurately and make the maximal precision during the relevancy calculation, the real-coded genetic algorithm (RCGA) is imported. All the three parameters a, b, c is real and limited between 0 and 1. For the realparameter problems, real-coded genetic algorithm performs better than some other encoding methods as binary coding and gray coding.

In order to increase the calculation efficiency and accuracy of sentence relevancycalculation and avoid the error brought by the interference information in sentences, pruning the dependency tree and extracting the keywords are needed.

3.1 Dependency Tree Pruning Rules

- For all the common sentences, the prime two levels of the dependence tree will be extracted and the rest of the branches will be cut off.
- In addition, if there is a comma or some other punctuation like that in the sentence, the prime two levels of the divided parts have to be extracted at the same time.
- If the pos of the words in the second level are auxiliary as "的", all their children words needed to be extracted to the second level while the original auxiliary father words should be deleted.
- If the pos of the words in the second level are verbs, all their children words in the third level under this verb need to be extracted and retained.

An example: 霍金是一位当代杰出人才。

(Hawking is an outstanding talent in the contemporary era.)

The dependency tree is shown in Fig.1 (a). And the pruning result is showed in Fig.1 (b) according to the rules.

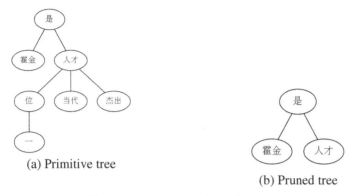

(a) Primitive tree

(b) Pruned tree

Fig. 1. An example of a dependency tree

3.2   Keywords Extraction Rule

All the word nodes in the pruned tree besides nouns, verbs and pronouns in the dependency tree of a sentence will be removed. While the nouns words include common nouns (n), azimuth nouns (nd), name nouns (nh), institution nouns (ni), site nouns (nl), time nouns (nt) and other proper nouns (nz).

Keywords of the example sentence above are '是', '霍金' and '人才'.

4   Experiments

Because there is no standard Chinese interactive text data set for sentence relevancy test now, a real world interactive text set from a QQ group named 'Linux group' is collected and labeled manually as our data set in the experiments. According to the experiments of words relevancy detection and sentence similarity, we use the collection of sentences to test our algorithm. While, 20 sentences, 10 relevant sentence pairs in the experiment are showed in Table 1 in the paper.

Most short sentences are with substandard language and incomplete semantic. Aiming at eliminating noise expressions in chat, a pre-process for interactive text is needed. After the pre-process, all the short sentences are standard and have complete semantic.

As is shown in Table 1, Column one in the table is the sentence id, column two describes the sentences and column three introduces the relevancy between the relevant pairs. From analysis of the results, we can see that, all the relevancies based on multi-features method are higher than that based on method without syntax feature. In conclusion, the multi-features method performs better than methods with no syntax feature in semantic relevancy measure of Chinese sentence. This sentence relevancy algorithm also performs well in topic detection method for interactive text [1] with 89.5% precision.

Table 1. Sentences relevancy results

Id	Sentences	Relevancy	
		No syntax feature	Multi-features
1	下午到底开会不，没有邮件通知啊。	0.44	0.50
2	同问，下午开会不？		
3	把教研室的同志都测量一下。	0.28	0.36
4	都测量一下，就发现林立的高于60了。		
5	你现在论文有我当年痛苦没？	0.39	0.41
6	想到论文就是这感觉了。		
7	明天小组9点开个小会，讨论年终奖的事情，请大家务必都准时参加。	0.37	0.42
8	我也去参加，分点年终奖。		
9	陈明请将ppclass系统设计与实现的工作总结尽快发我邮箱。	0.51	0.57
10	年终总结整合好的版本已经发到各位邮箱了，请查收。		
11	小明昨天和我发邮件了，说今天下午弄网络互连的事情。	0.80	0.74
12	我给博士发邮件他也没回。		
13	要不去看看有没有博士后？	0.17	0.20
14	心动啊，赶紧毕业，做博后去。		
15	大家找个时间去医院看看陈明吧。	0.30	0.34
16	可能他明天要住校医院。		
17	陈明住院了么？	0.49	0.55
18	今天他跟我说不用住院呀		
19	听他说不一定要做手术，医生说看看吃药有没有效果。	0.41	0.47
20	做手术的话，论文会被耽误的。		

5   Conclusions

This paper presents a novel method for detecting the relevancy of sentences in the Chinese interactive text. The method combines the information of the keyword-form feature, the semantic feature and the syntactic feature of a sentence. Experiments prove that the algorithm is more efficient in semantic relevancy measure of Chinese sentence. In addition, the proposed sentence relevancy algorithm also performs well in topic detection of Chinese interactive text.

The improvement of the sentence relevancy algorithm should be included in the further work, for example importing the interrogative as when, where, who and what, and adding the context relation to semantic relevancy measure of Chinese sentence. Besides, there is a lot of work need to do in the application of the common sense knowledge in interactive text, especially the un-complete short sentence in chat.

Acknowledgment. The research was supported in part by the National Science Foundation of China under Grant Nos.60825202, 60803079, 61070072, 61103160, the Fundamental Research Funds for the Central Universities under Grant No. xjj20100052, xjj20100057, the Doctoral Fund of Ministry of Education of China under Grant No. 20090201110060, Key Projects in the National Science & Technology Pillar Program during the Eleventh Five-Year Plan Period Grant No. 2008BAH26B02, 2009BAH51B02. The research was also supported by LTP which was explored by Information Retrieval Laboratory of Harbin Institute of Technology University.

References

1. Chen, Y., Yang, Y., Zhu, H.: Research on topic detection algorithm for interactive text in e-learning. Journal of Nanjing University of Posts and Telecommunications 31, 56–60 (2011)
2. Pingali, P., Jagarlamudi, J., Varma, V.: A relevance-based language modeling approach to duc (2005)
3. Zhang, Y., Xiong, F.: Text classification based on sentence correlation. Journal of University of Science and Technology of China 5, 540–545 (2006)
4. Sujian, L.: Research of relevancy between sentences based on semantic computa tion. Computer Engineering and Applications 7, 75–76 (2002)
5. Zhong, M., Liu, H., Zou, J.: The inter-sentence semantic relevancy degree calculation using the quantified correlation of words. Journal of Shandong University (Engineering Science) 5, 105–111 (2010)
6. Xu, Y., Fan, X., Zhang, F.: Semantic relevancy computing based on hownet. Journal of Beijing Institute of Technology 5, 411–414 (2005)
7. Zhang, K., Shen, X., Dong, F., Yu, J.: Semantic relevancy copmputing based on concept lattice. Journal of Zhengzhou University of Light Industry (Natural Science) 22, 178–181 (2007)
8. Jiaju, M., Yiming, Z., Yunqi, G.: Tongyici cilin, Shanghai (1983)
9. Li, B., Liu, T., Qin, B., Li, S.: Chinese sentence similarity computing based on semantic dependency relationship analysis. Application Research of Computers 12 (2003)

Author Index